Lecture Notes in Computer Science   9081

Commenced Publication in 1973
Founding and Former Series Editors:
Gerhard Goos, Juris Hartmanis, and Jan van Leeuwen

Editorial Board

More information about this series at http://www.springer.com/series/7408

Marco Gribaudo · Daniele Manini
Anne Remke (Eds.)

Analytical and Stochastic Modelling Techniques and Applications

22nd International Conference, ASMTA 2015
Albena, Bulgaria, May 26–29, 2015
Proceedings

 Springer

Editors
Marco Gribaudo
Politecnico di Milano
Milano
Italy

Anne Remke
Universität Münster
Münster
Germany

Daniele Manini
Università degli Studi di Torino
Torino
Italy

ISSN 0302-9743
Lecture Notes in Computer Science
ISBN 978-3-319-18578-1
DOI 10.1007/978-3-319-18579-8

ISSN 1611-3349 (electronic)

ISBN 978-3-319-18579-8 (eBook)

Library of Congress Control Number: 2015937539

LNCS Sublibrary: SL2 – Programming and Software Engineering

Printed on acid-free paper

Springer International Publishing AG Switzerland is part of Springer Science+Business Media (www.springer.com)

Preface

On behalf of the Chairs and the Technical Program Committee, it is our privilege to welcome all of you to ASMTA 2015, the 22nd International Conference on Analytical & Stochastic Modelling Techniques & Applications, that took place in the city of Albena (Varna) in Bulgaria.

The conference, thanks to its colocation with ECMS 2015 (the 29th European Conference on Modeling and Simulation), featured three distinguished keynote speakers who came from both the academia and the industry. Hans-Georg Zimmermann (Siemens Corporate Technology) delivered a talk on Causality versus Predictability in Neural Networks; Prof. Petko H. Petkov (Technical University of Sofia) presented studies on design and implementation of robust control laws; and Prof. Alexander H. Levis (George Mason University) introduced multi-formalism modeling of human organizations. We had 15 regular paper presentations composing the proceedings, which appeared in the Springer Verlag Lecture Notes in Computer Science (LNCS) series.

Many people have contributed to the realization of ASMTA 2015. First of all, our sincere thanks go to the authors who submitted their work to the conference. Special thanks go to the Technical Program Committee for their high-quality reviews and to the Program Chairs for their work in moderating the selection of the accepted papers. We are grateful for their help to the organizers of the colocated event ECMS 2015 for taking care of all the logistic and practical aspects. In particular, we wish to thank Martina M. Seidel for managing many aspects of the organization. We also wish to thank Dieter Fiems and Khalid Al-Begain for their support and guidance during the whole organization process.

We are grateful to our sponsor, The European Council for Modelling and Simulation. Finally, we would like to thank the EasyChair team and Springer for the editorial support of this conference series.

We do hope that you all find the conference a good opportunity for new experiences both from research and personal points of view. Welcome to Albena and enjoy the ASMTA 2015 program, the broader cooperation opportunity given by the colocation with a larger event, and the location itself!

March 2015

Marco Gribaudo
Daniele Manini
Anne Remke

Organization

Program Committee

Sergey Andreev	Tampere University of Technology, Finland
Jonatha Anselmi	Inria, France
Christel Baier	Technical University of Dresden, Germany
Simonetta Balsamo	Università Ca' Foscari di Venezia, Italy
Davide Cerotti	Politecnico di Milano, Italy
Lydia Chen	IBM Research lab Zurich, Switzerland
Antonis Economou	University of Athens, Section of Statistics and OR, Greece
Dieter Fiems	Ghent University, Belgium
Jean-Michel Fourneau	Universite de Versailles St Quentin, France
Marco Gribaudo	Politecnico di Milano, Italy
Yezekael Hayel	LIA, University of Avignon, France
András Horváth	University of Turin, Italy
Gábor Horváth	Budapest University of Technology and Economics, Hungary
Mauro Iacono	Seconda Università degli Studi di Napoli, Italy
Helen Karatza	Aristotle University of Thessaloniki, Greece
William Knottenbelt	Imperial College London, UK
Julia Kuhn	University of Queensland, Australia
Lasse Leskelä	Aalto University, Finnland
Daniele Manini	University of Turin, Italy
Andrea Marin	University of Venice, Italy
Jose Nino-Mora	Carlos III University of Madrid, Spain
Tuan Phung-Duc	Tokyo Institute of Technology, Japan
Pietro Piazzolla	Politecnico di Milano, Italy
Balakrishna J. Prabhu	LAAS-CNRS, France
Marie-Ange Remiche	University of Namur, Belgium
Anne Remke	Westfaelische Wilhelms Universitaet Muenster, Germany
Jacques Resing	Eindhoven University of Technology, The Netherlands
Marco Scarpa	University of Messina, Italy
Filippo Seracini	Microsoft, USA
Bruno Sericola	Inria, France
Janos Sztrik	University of Debrecen, Hungary
Miklos Telek	Budapest University of Technology and Economics, Hungary
Nigel Thomas	Newcastle University, UK

Dietmar Tutsch	University of Wuppertal, Germany
Benny Van Houdt	University of Antwerp, Belgium
Sabine Wittevrongel	Ghent University, Belgium
Verena Wolf	Saarland University, Germany
Katinka Wolter	Freie Universitaet zu Berlin, Germany
Alexander Zeifman	Vologda State University, Russia

Additional Reviewers

Angius, Alessio	Kovacs, Peter
Dei Rossi, Gian-Luca	Sandmann, Werner
Koops, David	Telek, Miklos

Contents

X   Contents

Optimal Analysis for M/G/1 Retrial Queue with Two-Way Communication

Amar Aissani[1]([⊠]) and Tuan Phung-Duc[2]

[1] RIIMA Laboratory, Department of Computer Science,
University of Sciences and Technology Houari Boumediene (USTHB),
BP 32, El Alia, 16 111 Bab-Ez-Zouar, Algiers, Algeria
aaissani@usthb.dz

[2] Tokyo Institute of Technology, Ookayama, Meguro-ku, Tokyo 152-8552, Japan
tuan@is.titech.ac.jp

Abstract. We consider an $M/G/1$ retrial queue with two types of calls: incoming calls (regular one's) and outgoing calls (which are made when the server is free). A blocked incoming call joins the orbit and retries for service after some random time while an outgoing call is made by the server after some random idle time. We assume that incoming and outgoing calls have random amount of works which are processed by the server at two distinct speeds. This assumption is suitable for evaluating the power consumption that depends on the speed of the server. We obtain the joint probability distribution of the server state and the number of requests in the orbit in terms of Laplace and z- transforms. From these transforms, we obtain some performance metrics of interest such as the probability that the server is idle or busy by an incoming (outgoing) call and the mean number of requests in orbit. We propose two optimization problems to find the optimal outgoing call rate and service speeds.

Keywords: Two way communication · QoS · Retrials · Performance-energy trade-off · Stationary distribution · Cost optimization

1 Introduction

Retrial queueing systems are interesting stochastic modeling tools, particularly in computer science. The blocked customer (or client, call, request) is allowed to repeat successively his attempt until the server is able to provide service. Otherwise, if the server is available, the arriving customer begins service immediately. These retrial queueing models have been used in modeling switching networks, wireless sensor networks, call centers and so on (see [3,4]).

The interested reader can find an extensive bibliography covering the 1990 − 1999 [3] and 2000 − 2009 periods [4], although the study of such problems begun with the queueing theory itself. Artalejo and Phung-Duc [5] exhibit some situations (e.g. call centers) in which the server not only serves incoming calls (regular one's), but also has a chance to make outgoing phone calls while they are not

© Springer International Publishing Switzerland 2015
M. Gribaudo et al. (Eds.): ASMTA 2015, LNCS 9081, pp. 1–14, 2015.
DOI: 10.1007/978-3-319-18579-8_1

engaged in conversation. This queueing feature is known as coupled switching or models of two way communication. Outgoing calls can also be seen as vacations or server breakdowns during which an incoming call cannot be served [1,2]. In Artalejo and Phung-Duc [5], a stationary analysis of the queue length process is carried out via embedded Markov chain and Markov renewal arguments. However, neither an analysis of optimality nor numerical results are presented by Artalejo and Phung-Duc [5].

In this paper, we consider the energetic version of the model of Artalejo and Phung-Duc [5] by assuming that each incoming or outgoing call requests a random amount of work which is processed by the server at a constant speed. This new interpretation is suitable for the computation of the power consumption which depends on the speed of the server [11]. The current model is suitable for a sensor node in wireless sensor networks. In wireless sensor network, a node collects data and sends the data to a sink. Packets that the node receives could be considered as incoming calls in the current model while data that the node sends to the sink can be seen as outgoing calls in our model. Because sensor node has a low capacity battery, the optimization of power consumption is an important issue. Thus, our optimization problems proposed in this paper may be useful in the design of such a system. Some other potential applications are described in [1]. We present an alternative solution based on supplementary variable method. We obtain the join stationary distribution of the number of calls in the orbit and the state of the server in terms of generating function and Laplace transform. These results are more general than those derived by Artalejo and Phung-Duc [5] in the sense that the latter is easily obtained form the former. We are able to derive some explicit formulae for the mean number of requests in the orbit as well as the state probability of the server. Based on these results, we propose two optimization problems to show the trade-off between power consumption and the performance of the system and find the optimal outgoing call rate and the service speeds of incoming and outgoing calls.

The rest of the paper is organized as follows. Section 2 describes the model in details while the analysis is presented in Section 3. Section 4 presents some performance measures and their numerical results. Section 5 proposes optimization problems with subject to the outgoing call rate and the speeds of the server. Finally, concluding remarks are presented in Section 6.

2   Model Description

We consider an $M_1, M_2/G_1, G_2/1$ retrial queue with two-way communication [5]. The flow generated by incoming calls forms a Poisson process with rate $\lambda > 0$. If such an incoming call finds the server free then his service begins immediately. During idle times, the server generates outgoing calls for other services.

We consider here the energetic interpretation of the service [1,2,9]. We consider that the service of a call can be interpreted as the realization of some work which needs some energy. Let S_n be the amount of work (speed or energy) to serve the n-th incoming call. We assume that the sequence $\{S_n, n \geq 1\}$ consists of independent identically distributed random variables with common distribution function $H_1(x)$, $H_1(0+) = 0$ and Laplace-Stieltjes transform $h_1(s)$, $Re(s) \geq 0$; the first and second order moments are denoted by h_{11}, h_{12}. Under this energetic interpretation, the server works with a power (or speed) $\gamma_1 > 0$ for an incoming (resp. $\gamma_2 > 0$ for an outgoing) call. If at time t an incoming primary or retrial (resp. outgoing) call arrives with a required energy x and if the server is free, then his service will be completed at time $t + x/\gamma_1$ (resp. at time $t + x/\gamma_2$), γ_1, $\gamma_2 > 0$. Now, if an arriving incoming call finds the server blocked by a service (of an incoming or an outgoing call), it becomes a source of secondary call and returns later in an exponentially distributed time with mean $1/\nu > 0$ to try again until it finds the server free; the collection of all secondary calls is called "orbit" (a sort of queue).

Now, if the server is free, then it generates an outgoing call in an exponentially distributed time with mean $1/\alpha$. The service times of the outgoing calls are independent with common probability distribution function $H_2(x)$, $H_2(0+) = 0$ and Laplace-Stieltjes transform $h_2(s)$, $Re(s) \geq 0$; the first and second order moments are denoted by h_{21} and h_{22}. The arrival flows of incoming and outgoing calls, service times and inter-retrial times are assumed to be mutually independent.

Consider the following random process $\zeta(t) = \{\alpha(t), R(t); \xi(t), t \geq 0\}$, where $\{R(t), t \geq 0\}$ is the number of customers in orbit at time t; $\alpha(t) = 0$, if the server is free at time t; $\alpha(t) = 1$ (resp. $\alpha(t) = 2$) if an incoming call (resp. an outgoing call) is in service at time t. The last component $\xi(t)$ is a positive real random variable: $\xi(t) = 0$, if $\alpha(t) = 0$; $\xi(t)$ is the residual service time if $\alpha(t) \neq 0$.

It is not difficult to show that the stochastic process $\{\zeta(t), t \geq 0\}$ is a Markovian process with piecewise linear paths which describes the evolution of the server state and the number of orbiting customers. We establish first the ergodicity condition for such a process, then we obtain its stationary probability distribution.

3   Analysis

3.1   Ergodicity Condition

Let

$$\rho = \lambda \frac{h_{11}}{\gamma_1}, \tag{1}$$

and

$$\sigma = \alpha \frac{h_{21}}{\gamma_2}. \tag{2}$$

The following theorem gives a condition for the existence of a stationary regime.

Theorem 1. *If*

$$\rho < 1, \tag{3}$$

then the stochastic process $\{\zeta(t), t \geq 0\}$ is ergodic and as a consequence there exists a unique stationary distribution. If $\rho > 1$, then the limiting probability distribution of $\{\zeta(t), t \geq 0\}$ tends to 0, and the underlying process tends to ∞.

Proof. The proof is similar to that of [2,5] with slight modifications.

3.2 Joint Distribution of the Server State and the Number of Calls in Orbit

In this section we derive the joint distribution of the server state and the number of customers in orbit in steady-state by its transform.

Under the assumption $\rho < 1$, the stochastic process $\{\zeta(t), t \geq 0\}$ is ergodic. As a consequence, the ergodic stationary probabilities

$$P_0(m) = \lim_{t \to \infty} P\{\alpha(t) = 0, R(t) = m\}, m \geq 0,$$

$$P_i(m, x) = \lim_{t \to \infty} P\{\alpha(t) = i, R(t) = m; \xi(t) < x\},$$

$$i = 1, 2, m \geq 0, x \geq 0,$$

are solutions of the following system of differential equations

$$(\lambda + \alpha + \nu m)P_0(m) = \gamma_1 \frac{dP_1(m, 0)}{dx} + \gamma_2 \frac{dP_2(m, 0)}{dx}, m \geq 0,$$

$$\lambda P_1(m, x) = \gamma_1 \frac{dP_1(m, x)}{dx} - \gamma_1 \frac{dP_1(m, 0)}{dx} + \lambda(1 - \delta_{0m})P_1(m - 1, x) +$$

$$+ \lambda P_0(m)H_1(x) + \nu(m + 1)P_0(m + 1)H_1(x), m \geq 0, x \geq 0,$$

$$\lambda P_2(m, x) = \gamma_2 \frac{dP_2(m, x)}{dx} - \gamma_2 \frac{dP_2(m, 0)}{dx} + \lambda(1 - \delta_{0m})P_2(m - 1, x) +$$

$$+ \alpha P_0(m)H_2(x), m \geq 0, x \geq 0,$$

where δ_{ij} is the Kronecker function. We introduce the partial generating functions in z,

$$Q_0(z) = \sum_{m=0}^{\infty} z^m P_0(m),$$

$$Q_i(z, x) = \sum_{m=0}^{\infty} z^m P_i(m, x), \quad i = 1, 2.$$

Applying these transforms to the previous system, we obtain

$$(\lambda + \alpha)Q_0(z) + \nu z \frac{dQ_0(z)}{dz} = \gamma_1 \frac{\partial Q_1(z, 0)}{\partial x} + \gamma_2 \frac{\partial Q_2(z, 0)}{\partial x}, \tag{4}$$

$$(\lambda - \lambda z)Q_1(z, x) = \gamma_1 \frac{\partial Q_1(z, x)}{\partial x} - \gamma_1 \frac{\partial Q_1(z, 0)}{\partial x} + \lambda Q_0(z)H_1(x) + \nu \frac{dQ_0(z)}{dz}H_1(x), \tag{5}$$

$$(\lambda - \lambda z)Q_2(z,x) = \gamma_2 \frac{\partial Q_2(z,x)}{\partial x} - \gamma_2 \frac{\partial Q_2(z,0)}{\partial x} + \alpha Q_0(z)H_2(x). \qquad (6)$$

We apply now the Laplace transform to the second argument of the obtained partial generating functions in equations (5) and (6) and we get

$$s(\gamma_1 s - \lambda + \lambda z)f_1(z,s) = \gamma_1 \frac{\partial Q_1(z,0)}{\partial x} - \left(\lambda Q_0(z) + \nu \frac{dQ_0(z)}{dz}\right)h_1(s), \qquad (7)$$

$$s(\gamma_2 s - \lambda + \lambda z)f_2(z,s) = \gamma_2 \frac{\partial Q_2(z,0)}{\partial x} - \alpha Q_0(z)h_2(s), \qquad (8)$$

where $f_i(z,s)$ denotes the Laplace transform of $Q_i(z,x)$ $(i = 1,2)$, i.e.,

$$f_i(z,s) = \int_0^\infty e^{-sx} Q_i(z,x)dx, \qquad i = 1,2.$$

The unknown functions $\frac{\partial Q_i(z,0)}{\partial x}$, $i = 1,2$ can be determined as usual by using the fact that the functions $Q_i(z,0)$ are analytical functions in the domain $|z| \leq 1$. Consider for example equation (8) of the previous system of equations. Since $Q_2(z,0)$ is analytic in the domain $|z| \leq 1$, and since the left hand is equal to zero for $s = (\lambda - \lambda z)/\gamma_2$, then the right hand side must also be zero at this point. As a result, we have the first condition

$$\frac{\partial Q_2(z,0)}{\partial x} = \frac{\alpha}{\gamma_2} Q_0(z)h_2\left(\frac{\lambda - \lambda z}{\gamma_2}\right). \qquad (9)$$

Similarly, $Q_1(z,0)$ is analytic in the domain $|z| \leq 1$, and since the left right hand is equal to zero for $s = (\lambda - \lambda z)/\gamma_1$, then the right hand side must also be zero at this point. As a result, we also have

$$\frac{\partial Q_1(z,0)}{\partial x} = \frac{1}{\gamma_1}\left(\lambda Q_0(z) + \nu \frac{dQ_0(z)}{dz}\right)h_1\left(\frac{\lambda - \lambda z}{\gamma_1}\right). \qquad (10)$$

Substituting now (9)-(10) in (7)-(8), we obtain the functions $f_i(z,s)$ $(i = 1,2)$ under the following explicit form

$$f_1(z,s) = \frac{\left[h_1\left(\frac{\lambda - \lambda z}{\gamma_1}\right) - h_1(s)\right]\left[\lambda Q_0(z) + \nu \frac{dQ_0(z)}{dz}\right]}{s(\gamma_1 s - \lambda + \lambda z)}, \qquad (11)$$

$$f_2(z,s) = \alpha \frac{h_2\left(\frac{\lambda - \lambda z}{\gamma_2}\right) - h_2(s)}{s(\gamma_2 s - \lambda + \lambda z)} Q_0(z). \qquad (12)$$

Now, substitution of (9)-(10) in equation (4) gives

$$(\lambda + \alpha)Q_0(z) + \nu z \frac{dQ_0(z)}{dz} = \left(\lambda Q_0(z) + \nu \frac{dQ_0(z)}{dz}\right)h_1\left(\frac{\lambda - \lambda z}{\gamma_1}\right) + \alpha Q_0(z)h_2\left(\frac{\lambda - \lambda z}{\gamma_2}\right), \qquad (13)$$

or equivalently,

$$\left(\lambda[1 - h_1\left(\frac{\lambda - \lambda z}{\gamma_1}\right)] + \alpha[1 - h_2\left(\frac{\lambda - \lambda z}{\gamma_2}\right)] \right) Q_0(z) = \nu[h_1\left(\frac{\lambda - \lambda z}{\gamma_1}\right) - z]\frac{dQ_0(z)}{dz}. \tag{14}$$

The solution of this homogeneous ordinary differential equation is of the form

$$Q_0(z) = k_0 exp\left(\frac{\lambda}{\nu}\int_0^z \frac{1 - h_1\left(\frac{\lambda - \lambda y}{\gamma_1}\right)}{h_1\left(\frac{\lambda - \lambda y}{\gamma_1}\right) - y}dy\right) \times exp\left(\frac{\alpha}{\nu}\int_0^z \frac{1 - h_2\left(\frac{\lambda - \lambda y}{\gamma_2}\right)}{h_1\left(\frac{\lambda - \lambda y}{\gamma_1}\right) - y}dy\right). \tag{15}$$

The constant k_0 can be determined using the normalization condition and it will be done in the next section.

So, we have the following result.

Theorem 2. *If the condition $\rho < 1$ is fulfilled, then the joint distribution of the server state, orbit size and residual work is given by it's transform*

$$f_1(z, s) = \frac{\left[h_1\left(\frac{\lambda - \lambda z}{\gamma_1}\right) - h_1(s)\right]\left(\lambda(1 - z) + \alpha[1 - h_2\left(\frac{\lambda - \lambda z}{\gamma_2}\right)]\right)}{s(\gamma_1 s - \lambda + \lambda z)[h_1\left(\frac{\lambda - \lambda z}{\gamma_1}\right) - z]}Q_0(z), \tag{16}$$

$$f_2(z, s) = \alpha\frac{h_2\left(\frac{\lambda - \lambda z}{\gamma_2}\right) - h_2(s)}{s(\gamma_2 s - \lambda + \lambda z)}Q_0(z) \tag{17}$$

where the function $Q_0(z)$ is given in Theorem 3.

Proof. Using Tauberian theorem in formula (11), we get

$$Q_1(z, \infty) = lim_{x \to \infty}Q_1(z, x) = lim_{s \to 0+}sf_1(z, s) =$$

$$= \frac{1 - h_1\left(\frac{\lambda - \lambda z}{\gamma_1}\right)}{\lambda - \lambda z}\left(\lambda Q_0(z) + \nu\frac{dQ_0(z)}{dz}\right). \tag{18}$$

Similarly, we obtain from (12)

$$Q_2(z, \infty) = \alpha\frac{1 - h_2\left(\frac{\lambda - \lambda z}{\gamma_2}\right)}{\lambda - \lambda z}Q_0(z). \tag{19}$$

Using equation (15) we have

$$\lambda Q_0(z) + \nu\frac{dQ_0(z)}{dz} = \frac{\lambda(1 - z) + \alpha[1 - h_2\left(\frac{\lambda - \lambda z}{\gamma_2}\right)]}{h_1\left(\frac{\lambda - \lambda z}{\gamma_1}\right) - z}Q_0(z). \tag{20}$$

Hence, by substituting (20) into (18), we obtain

$$Q_1(z, \infty) = \frac{1 - h_1\left(\frac{\lambda - \lambda z}{\gamma_1}\right)}{h_1\left(\frac{\lambda - \lambda z}{\gamma_1}\right) - z}\left[1 + \frac{\alpha[1 - h_2\left(\frac{\lambda - \lambda z}{\gamma_2}\right)]}{\lambda - \lambda z}\right]Q_0(z). \tag{21}$$

The relation (16) is obtained by substituting (18) into (11) and equation (17) is just (12).

3.3   Distribution of the Number of Calls in Orbit

Let $q_m = P\{R(t) = m\}, m = 0, 1, \dots$ be the distribution of the number of calls in orbit.

It is easy to see that

$$q_m = P_0(m) + P_1(m, \infty) + P_2(m, \infty) \tag{22}$$

The generating function of this distribution is

$$Q(z) = \sum_{m=0}^{\infty} q_m z^m = Q_0(z) + Q_1(z, \infty) + Q_2(z, \infty)$$

Theorem 3. *If $\rho = \frac{\lambda h_{11}}{\gamma_1} < 1$, then the generating function of the number of calls in orbit is given by*

$$Q(z) = \frac{\lambda - \lambda z + \alpha\left[1 - h_2\left(\frac{\lambda - \lambda z}{\gamma_2}\right)\right]}{\lambda\left[h_1\left(\frac{\lambda - \lambda z}{\gamma_1}\right) - z\right]} Q_0(z), \tag{23}$$

where

$$Q_0(z) = \frac{1 - \rho}{1 + \alpha h_{21}/\gamma_2} exp\left(\frac{\lambda}{\nu} \int_z^1 \frac{1 - h_1\left(\frac{\lambda - \lambda y}{\gamma_1}\right)}{h_1\left(\frac{\lambda - \lambda y}{\gamma_1}\right) - y} dy\right) \times exp\left(\frac{\alpha}{\nu} \int_z^1 \frac{1 - h_2\left(\frac{\lambda - \lambda y}{\gamma_2}\right)}{h_1\left(\frac{\lambda - \lambda y}{\gamma_1}\right) - y} dy\right). \tag{24}$$

Proof. Taking into account (15), (17) and (19), we obtain after some algebra expression (23). From this formula we can obtain an explicit expression of the constant k_0 by using the normalization condition $Q(1) = 1$. Note that when taking $z = 1$ in (23), we obtain an indeterminate form $0/0$. This difficulty can be overcome by using twice l'Hospital's rule.

Remark 1. Note that if the energy $\gamma_1 = \gamma_2 = 1$, then we obtain the results by Artalejo and Phung-Duc [5] for the temporal model with two-way communication.

Remark 2. We note that the ergodicity condition $\rho < 1$ appears here since the constant k_0 must be strictly positive.

4   Some Performance Metrics and Numerical Results

In this section we show how the results of the previous sections can be helpful to derive several performance metrics of interest.

4.1   Performance Metrics

From (15), (17), (23) and (24) we can obtain some performance metrics of interest.

- The probability that the server is idle:

$$P_0 = Q_0(1) = Q_0(z) \mid_{z=1} = \frac{1 - \lambda h_{11}/\gamma_1}{1 + \alpha h_{21}/\gamma_2}. \tag{25}$$

- The probability that the server is busy by incoming service:

$$P_1 = Q_1(1, \infty) = \frac{\lambda}{\gamma_1} h_{11}. \tag{26}$$

- The probability that the server is busy by outgoing service:

$$P_2 = Q_2(1, \infty) = \frac{(1 - \lambda h_{11}/\gamma_1)(\alpha h_{21}/\gamma_2)}{1 + \alpha h_{21}/\gamma_2}. \tag{27}$$

- The mean number of calls in orbit:

$$N_1 = Q'(1) = \frac{dQ(z)}{dz} \mid_{z=1} = \frac{\lambda \alpha h_{22}/\gamma_2^2}{2(1 + \alpha h_{21}/\gamma_2)} + \frac{\lambda^2 h_{12}/\gamma_1^2}{2(1 - \lambda h_{11}/\gamma_1)} +$$
$$+ \frac{\lambda(\lambda h_{11}/\gamma_1 + \alpha h_{21}/\gamma_2)}{\nu(1 - \lambda h_{11}/\gamma_1)} \tag{28}$$

- The mean number of calls in orbit when the server is idle:

$$M_1 = Q_0'(1) = \frac{dQ_0(z)}{dz} \mid_{z=1} = \frac{\lambda}{\nu} \frac{\lambda h_{11}/\gamma_1 + \alpha h_{21}/\gamma_2}{1 + (\alpha h_{21})/\gamma_2}. \tag{29}$$

Remark 3. We note that the idle server probability is independent of the retrial rate ν. Indeed, this probability depends only on $\sigma = \alpha h_{21}/\gamma_2$ (the traffic intensity of outgoing calls) which is clearly independent of retrial rate and $\rho = \lambda h_{11}/\gamma_1$ (the traffic intensity of incoming calls) which is also independent of retrial rate for linear retrial policy (in opposition to constant retrial policy). On the other hand the ergodicity condition is the same as for the energetic $M/G/1$ FIFO [9] queue since each incoming call must be served.

4.2   Numerical Examples

This subsection is devoted to the presentation of some examples and numerical illustrations showing how to exploit the results of the previous sections. The objective is to indicate to a practitioner how he can observe the influence of given parameters upon given performance measures of interest.

Figure 1(i) illustrates the effect of service speed of outgoing calls γ_2 on the idle server probability P_0 for different values of γ_1: 5, 10 and 20. We note that P_0 increases when both outgoing rate γ_1 and γ_2 increase. Figure 1(ii) shows

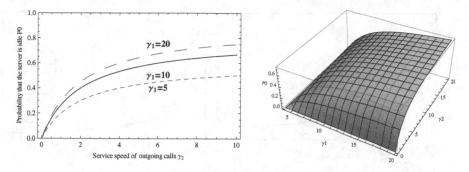

Fig. 1. (i) Effect of speed of outgoing calls γ_2 on the idle server probability P_0. (ii) Effect of γ_1 and γ_2 on the idle server probability P_0.

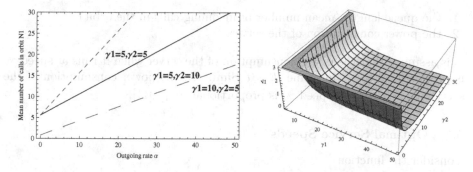

Fig. 2. (i) Effect of outgoing rate α for different values of γ_1 and γ_2 on N_1. (ii) Effect of γ_1 and γ_2 on N_1.

the same evolution on a $3D$ graph. Some parameters are fixed as follow $\lambda = 1$, $\alpha = 2, h_{11} = 2, h_{21} = 1, \nu = 2$.

In Figure $2(i)$ we show the effect of outgoing rate on the mean number of calls in orbit N_1 for fixed values of service speeds for both incoming and outgoing calls $(\gamma_1, \gamma_2) = (5,5), (5,10)$ and $(10,5)$. As expected, the mean number of calls in orbit linearly increases with increasing of the outgoing rate α. Figure $2(ii)$ shows that N_1 decreases when both γ_1 and γ_2 increase.

Finally, Figures $3(i)$ and $3(ii)$ illustrates the effect of retrial rate ν on the mean number of calls in orbit N_1 and mean number of calls in orbit when the server idle M_1 for different values of $(\gamma_1, \gamma_2) = (5,5)$ (short dashed line), $(5,10)$ (gray level line), $(10,5)$ (long dashed line). As expected both N_1 and M_1 decreases with decreasing of ν.

5 Optimization Problems

The energetic interpretation considered here leads to a new optimization problem trying to minimize the total cost of the system. In fact, there are two factors that we want to minimize.

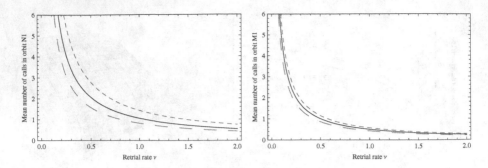

Fig. 3. (i) Effect of retrial rate ν on N_1. (ii) Effect of retrial rate ν on M_1.

1. the queue length (mean number of incoming calls in the orbit)
2. the power consumption of the server.

It is assumed that the power consumption of the server when it runs at speed γ_1 is proportional to γ_1^β for some $\beta > 0$. Similarly, the power consumption of the server when it runs at speed γ_2 is proportional to γ_2^β [11].

5.1 Optimal Service Speeds

Consider the function

$$F(\gamma_1, \gamma_2) = N_1 + \frac{1}{\delta} E(S), \tag{30}$$

which represents the total cost of the system, where $E(S) = C \times P_0 + P_1 \times \gamma_1^\beta + P_2 \times \gamma_2^\beta$ and δ is the relative importance on power consumption and queue length (QoS). So, we can suggest the following optimization problem.

$$min_{\gamma_1>0,\gamma_2>0} F(\gamma_1, \gamma_2). \tag{31}$$

In the following sections, we provide several numerical experiments showing some non trivial values of γ_1 and γ_2 which minimize the total cost of the system given by expression (31), where the constant P_0, P_1, P_2 and N_1 are given by (25)-(28).

5.2 Optimal Idle Interval

We now consider the optimization problem subject to the mean idle interval. In particular, we find the optimal α which minimizes the cost function while keeping all other parameters constant.

This optimization problem is useful in the situation where we need to determine the timing to make an outgoing call. A large mean idle interval $(1/\alpha)$ reduces the mean queue length. At the same time it increases the idle probability resulting in the increase of useless energy consumption. It should be noted that idle server consumes energy but does not process jobs.

Because h_{21}, γ_2 are constant, minimizing subject to α is equivalent to minimizing subject to σ which is the traffic intensity of outgoing calls. Taking the derivative with respect to σ yields

$$\frac{dF(\gamma_1, \gamma_2)}{d\sigma} = -\frac{A}{(1+\sigma)^2} + B,$$

where A and B are given by

$$A = \frac{(1-\rho)C}{\delta} - \frac{(1-\rho)\gamma_2^\beta}{\delta} - \frac{\lambda h_{22}}{2\gamma_2 h_{21}}, \qquad B = \frac{\lambda}{\nu(1-\rho)}.$$

If $A \leq B$, we have

$$\frac{dF(\gamma_1, \gamma_2)}{d\sigma} > 0$$

implying that the cost function is minimized at $\sigma = 0$. If $A > B$, the cost function is minimized at

$$\sigma^* = \sqrt{\frac{A}{B} - 1}.$$

5.3   Examples and Numerical Illustrations

In this section, we illustrate the effect of different parameters on the cost function in order to show performance-energy trade-off.

The parameters are fixed as follow $\lambda = 2$, $\alpha = 2$, $h_{11} = 2$, $h_{12} = 1$, $h_{21} = 1$, $h_{22} = 2$, $\nu = 2$, $C = 1$, $\beta = 2$. Now we want to illustrate the effect of γ_1 and γ_2 on the cost function.

We present in Table 1 the optimal values of γ_1 and γ_2 and the corresponding minimal cost for different values of δ. We see that the minimal value of the cost function decreases when δ increases.

In Figures $4(i)$, $4(ii)$ and $5(i)$ we represent the effect of γ_1 on the cost function $F(\gamma_1, \gamma_2)$ for different special cases ($\delta = 0.1$).

- $\gamma_1 = 10 \times \gamma_2$: short dashed line,
- $\gamma_1 = \gamma_2$: Gray level line,
- $\gamma_1 = 0.1 \times \gamma_1$: long dashed line.

Note that due to the ergodicity condition, the range of γ_1 is respectively $\gamma_1 > 4$, $\gamma_1 > 0.4$ and $\gamma_1 > 40$. Figure $5(ii)$ shows all graphs together. We note that the Figure $4(i)$ is masked by Figure $5(i)$ in the left up corner on the graph due to scaling.

Figures 6 and 7 show the effect of the retrial rate on the optimal σ^* for different values of C and δ. The optimal σ^* increases when both δ and retrial rate increase.

Table 1. Optimal values of γ_1^*, γ_2^* and optimal cost $F(\gamma_1^*, \gamma_2^*)$ for different values of δ: $C = 1$, $\nu = 2$.

δ	Optimal value γ_1^*,	Optimal value γ_2^*	Minimum cost $F * (\gamma_1^*, \gamma_2^*)$
0.1	4.03353	4.10481×10^6	16268.3
0.5	4.07498	4.93492×10^6	3319.91
1	4.10603	4.51583×10^6	1684.76
5	4.23706	4.39855×10^6	357.841
10	4.33521	2.58957×10^6	186.725
15	4.41049	1.20018×10^6	128.468
20	4.47393	883739	98.8672
30	4.58029	650642	68.7203
50	4.7488	350133	43.8979
100	5.05787	193073	24.3886
300	5.8268	51323.7	10.1471
500	6.35329	35950.9	6.91697
1000	7.31537	15547.9	4.21571
5000	11.3242	2887.86	1.47686
10000	14.2982	1633.86	0.97487

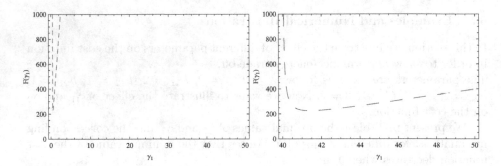

Fig. 4. (i) Cost function versus γ_1 for $\gamma_1 = 10 \times \gamma_2$: $C = 1$, $\nu = 2$. (ii) Cost function versus γ_1 for $\gamma_1 = 0.1 \times \gamma_2$: $C = 1$, $\nu = 2$.

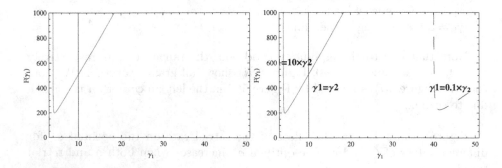

Fig. 5. (i) Cost function versus γ_1 for $\gamma_1 = \gamma_2$: $C = 1$, $\nu = 2$. (ii) All graphs together.

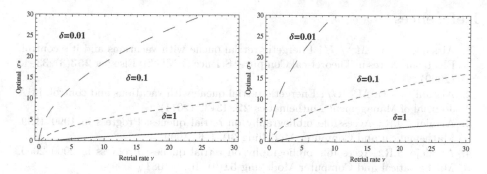

Fig. 6. (i) Effect of δ and retrial rate ν on the optimal value σ^*: $C = 20$, $\gamma_1 = 5$, $\gamma_2 = 1$ (ii) Effect of δ and retrial rate ν on the optimal value σ^*: $C = 50$, $\gamma_1 = 5$, $\gamma_2 = 1$.

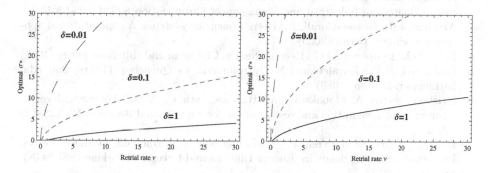

Fig. 7. (i) Effect of δ and retrial rate ν on the optimal value σ^*: $C = 30$, $\gamma_1 = 10$, $\gamma_2 = 5$ (ii) Effect of δ and retrial rate ν on the optimal value σ^*: $C = 50$. $\gamma_1 = 10$, $\gamma_2 = 5$.

6   Conclusion

In this work, we have provided an analysis of the energetic version of the $M/G/1$ retrial queue with two-way communication studied by Artalejo and Phung-Duc [5]. In particular, we have made the conjecture that this energetic interpretation leads to a new optimization problem trying to minimize the total cost of the system. This conjecture was verified experimentally by showing the evolution of the cost function for different values of the parameters. The next step will be to provide an analytical proof of this conjecture. It will be interesting also to study the influence of reliability and/or vacations.

Acknowledgments. This work was supported in parts by the Algerian ministry of high education and scientific research through grants B00220100046 and $B * 00220140064$. It has been initiated during a visit of the first author at Tokyo Institute of Technology (Japan) on behalf of 10th Workshop on Retrial Queues July $24 - 26$, 2014. The authors would like to thanks the anonymous referees for pertinent and helpful comments that allow to improve the quality of the paper.

References

1. Aissani, A.: An $M^X/G/1$ Energetic retrial queue with vacations and it's control. Electronic Notes in Theoretical Computer Science (ENTCS) Elsevier **253**(3), 33–44 (2009)
2. Aissani, A.: An $M^X/G/1$ Energetic retrial queue with vacations and control. IMA Journal of Management Mathematics **22**, 13–32 (2011)
3. Artalejo, J.R.: Accessible bibliography on retrial queues: Progress in 1990–1999. Mathematical and Computer Modelling **30**, 1–6 (1999)
4. Artalejo, J.R.: Accessible bibliography on retrial queues: Progress in 2000–2009. Mathematical and Computer Modelling **51**(9), 1071–1081 (2009)
5. Artalejo, J.R., Phung-Duc, T.: Single server retrial queues with two way communication. Applied Mathematical Modelling **37**, 1811–1822 (2013)
6. Artalejo, J.R., Phung-Duc, T.: Markovian retrial queues with two way communication. Journal of Industrial and Management Optimization **8**, 781–806 (2012)
7. Artalejo, J.R., Gomez-Corall, A.: Retrial queueing systems: A computational approach. Springer, Berlin (2008)
8. Falin, G.I., Templeton, J.G.: Retrial Queues. Chapman and Hill, New Jersey (1997)
9. Gnedenko, B.V., Kovalenko, I.N.: Introduction to Queueing Theory, 2nd edn. Birkhauser, Boston (1989)
10. Krishnamoorthy, A., Gobakamarand, B., Viswanath, C.: A retrial queue with server interruptions, resumption and restart of service. Operational Research **12**, 133–149 (2012)
11. Lu, X., Aalto, S., Lassila, P.: Performance-energy trade-off in data centers: Impact of switching delay. In: Energy Efficient and Green Networking (SSEEGN), pp. 50–55 (2013)

Use of Flow Equivalent Servers in the Transient Analysis of Product Form Queuing Networks

Alessio Angius[1], András Horváth[1(✉)], Sami M. Halawani[2], Omar Barukab[2], Ab Rahman Ahmad[2], and Gianfranco Balbo[1,2]

[1] Dipartimento di Informatica, Università di Torino, Turin, Italy
{angius,horvath,balbo}@di.unito.it
[2] Faculty of Computing and Information Technology, King Abdulaziz University,
Rabigh Branch, Rabigh, Kingdom of Saudi Arabia
{halawani,obarukab,abinahmad}@kau.edu.sa

Abstract. In this paper we deal with approximate transient analysis of Product Form Queuing Networks. In particular, we exploit the idea of flow equivalence to reduce the size of the model. It is well-known that flow equivalent servers lead to exact steady state solution in many cases. Our goal is to investigate the applicability of flow equivalence to transient analysis. We show that exact results can be obtained even in the transient phase, but the definition of the equivalent server requires the analysis of the whole original network. We propose thus to use approximate aggregate servers whose characterization demands much less computation. Specifically, the characterization corresponds to the steady state equivalent server of the stations that we aim to aggregate and thus can be achieved by analyzing the involved stations in isolation. This way, approximations can be derived for any queuing network, but the precision of the results depends heavily on the topology and on the parameters of the model. We illustrate the approach on numerical examples and briefly discuss a set of criteria to identify the cases when it leads to satisfactory approximation.

1 Introduction

Queuing networks are widely used for the design and analysis of Discrete Event Dynamic Systems [18]. The complexities of real systems translate in the complexities of their models, thus making the understanding of their properties and behaviors a computationally difficult task. *Product Form Queuing Networks* (PFQNs) [6] are a restricted class of queuing networks with special features which make them interesting for practical purposes. Many applications of PFQN can be found in the literature for the analysis of Computer, Communication, and Manufacturing Systems. The derivation of efficient computational algorithms for the solution of their mathematical representations [10] has made them very popular because of the possibility of analyzing very large models with limited computational effort. Still, given the ever increasing complexity of real systems, simplification techniques with which an entire sub-model (comprising many nodes -or

M. Gribaudo et al. (Eds.): ASMTA 2015, LNCS 9081, pp. 15–29, 2015.
DOI: 10.1007/978-3-319-18579-8_2

servers- of the original network) can be identified within the model, analyzed in isolation, and replaced with a single *flow equivalent server* are very attractive because they reduce the computational complexity of the analysis of the whole model. The characterization of the flow equivalent server and the exactness of the substitution rely on the characteristics of the product form solution of the original model [5,11,17]. Product form is a property of the stationary distribution of PFQN and holds only in very specific cases in the transient phase [8]. As a consequence, the transient analysis of real systems gains little advantages when these same systems can be modeled with PFQNs. Moreover, little is known about the quality (approximation level) of the transient solution of a model computed when subsystems are replaced by their flow equivalent counterparts.

In this paper we investigate the application of flow-equivalence for transient analysis. We show that the *transient* solution of a model reduced by the concept of steady state flow equivalence may yield a good approximation of the transient behavior of the original model. We limit ourselves to an approximation because, excluding very special cases, it appears unrealistic to devise an exact approach that reduces the complexity of the transient analysis as it happens for the steady state analysis of PFQNs. On the basis of these preliminary observations, we suggest certain conditions that must be satisfied by the original model for the method to apply in a satisfactory manner.

Transient analysis is a computationally difficult problem when the state space is large (more than about 10^7 states) and not many techniques have been proposed to solve it. Apart of the simplest cases [8], only approximate and simulation based techniques are viable. Among the approximate approaches we have moment closure techniques [19] and fluid approximations [14]. Methods based on aggregation have also been developed, see, for example, [7]. Fewer techniques maintain the original state space of the model and, as a consequence, allow one to calculate also distributions. Among these, memory efficient approaches have been proposed based on assuming that the transient probabilities are in a special form, like product form [1], partial product form [20], or quasi product form [3]. A decomposition based technique has been proposed instead in [2].

The paper is organized as follows. In Section 2 we introduce the model and summarize the methods used to compute the stationary distribution of PFQNs. In Section 3 we describe the concept of *flow equivalent server* and show how to restructure the model to accommodate the flow equivalent server. In Section 4 we discuss the difficulties that arise when using flow equivalence in a finite time domain. In Section 5 we discuss the quality of the approximation by applying the method to (relatively) large networks exhibiting different characteristics and draw some general indications concerning both the cases in which the method performs well as well as those where the results are not satisfactory. Finally, in Section 6 we draw conclusions.

2   Model

We consider a closed network of M queues, $\mathcal{N} = \{s_1, s_2, \ldots, s_M\}$, with a fixed number N of statistically identical jobs circulating through the network at all times (single class network). Jobs (or customers) get service at the stations and move from one station to another according to pre-fixed routing probabilities denoted by $r_{i,j}$ with $i, j \in \{1, 2, ..., M\}$ and globally represented by the routing matrix $\boldsymbol{R}$. Service times have exponential distributions which may be *load dependent* (i.e. functions of the number jobs at the station) and the service discipline is FIFO. The service intensity (rate) of queue i in the presence of n jobs at the station is denoted by $\mu_i(n)$.

A given state of the model is a vector $\boldsymbol{n} = [n_1, ..., n_M]$ where n_i denotes the number of jobs at station i and $\sum_{i=1}^{M} n_i = N$. Accordingly, the state space is defined as

$$ \mathcal{S}(N, M) = \left\{ \boldsymbol{n} = [n_1, ..., n_M] \,\middle|\, n_i \geq 0, i = 1, ..., M; \sum_{i=1}^{M} n_i = N \right\} \tag{1} $$

and its cardinality is $|\mathcal{S}(N, M)| = \binom{N+M-1}{M-1}$.

The number of jobs at station i at time t is denoted by $n_i(t)$. The probability that the system is in state $\boldsymbol{n}$ at time t is denoted by $\pi_{\boldsymbol{n}}(t)$. The probabilities of the whole state space are conveniently collected in a vector $\boldsymbol{\pi}(t) = [\pi_{\boldsymbol{n}}(t)]_{\boldsymbol{n} \in \mathcal{S}(N,M)}$.

It has been proved by Gordon and Newell [16] that the equilibrium distribution of customers in a closed PFQN of this type is given by

$$ \pi_{\boldsymbol{n}} = \frac{1}{G} \prod_{i=1}^{M} f_i(n_i) \tag{2} $$

where G is a normalizing constant defined so that we have $\sum \pi_{\boldsymbol{n}} = 1$, and the function f_i (often called *service function* of station i) is defined as

$$ f_i(n) = \begin{cases} 1 & n = 0 \\ \frac{V_i}{\mu_i(n)} f_i(n-1) & n \geq 1 \end{cases} \tag{3} $$

being $\boldsymbol{V} = [V_1, ..., V_M]$ a real positive solution of the eigenvector-like equation

$$ \boldsymbol{V} = \boldsymbol{V} \boldsymbol{R} \tag{4} $$

The direct computation of the normalization constant G is of exponential complexity, but, due to the pioneering work of Buzen [10], computationally efficient algorithms have been devised [13] which obtain the desired quantity in polynomial time. One such method, called the convolution algorithm, is based on the definition of an auxiliary function $g(n, m)$ that represents the normalization constant computed for the network comprising the first $m (\leq M)$ stations of the original network and in which only $n (\leq N)$ customers are present. This quantity can be computed by a recursive method based on the convolution-type summation

$$ g(n, m) = \sum_{k=0}^{n} f_m(k) \, g(n-k, m-1) \tag{5} $$

The normalization constant for the whole network with M jobs is given then by $G = g(N, M)$, with $g(n, 1) = f_1(n)$.

The method provides expressions for many standard performance indices of the network that can be defined in terms of values obtained during the computation of G [10]. An important result, that was also originally derived by Buzen in [10], concerns the marginal distribution of the customers at the ith station. For the purposes of this paper, we can focus our attention on the distribution of the customers in the last station of the network. The probability of having h jobs at the Mth station can be computed without considering all states of the network as simply as

$$P_M(h) = f_M(h) \frac{g(N - h, M - 1)}{g(N, M)} \tag{6}$$

3   Flow-Equivalent Aggregation

On the basis of the results recalled in the previous section, we can now define an *equivalent server*. Consider a PFQN $\mathcal{N}$ defined as before and let $\mathcal{K} = \{s_1, s_2, \ldots, s_K\}, s_i \in \{1, \ldots, M\}$, be a subset of its stations. Let $\mathcal{K}'$ be the complementary subset of stations of $\mathcal{K}$ so that $\mathcal{K} \cap \mathcal{K}' = \emptyset$ and $\mathcal{K} \cup \mathcal{K}' = \mathcal{N}$. For sake of simplicity, in the rest of this paper, unless when specified differently, we assume that the stations of the network are ordered so that those of sub-network $\mathcal{K}$ are the first K stations of the network $\mathcal{N}$ that in the rest of the paper will be often referred to as the *aggregated stations*. The goal of the aggregation step is to characterize the behavior of the sub-model $\mathcal{K}$ in order to replace it with an equivalent server that will then interact with the other sub-model $\mathcal{K}'$ without affecting its behavior. This characterization of the equivalent server is performed with a *controlled experiment* [11,12,15] which corresponds to solving the sub-model $\mathcal{K}$ in isolation under a fixed load $H = 1, \ldots, N$. For this purpose let us assume $\mathcal{N}'$ to be a queuing network having the same topology of $\mathcal{N}$, but such that the stations not in $\mathcal{K}$ have null service times (they are often referred to as "short-circuited"). Denoting by π' the stationary distribution of the customers in the network $\mathcal{N}'$, and with $\boldsymbol{X}$ a generic state of $\mathcal{N}'$

$$\pi'_{\boldsymbol{n}}(H) = Pr\{X_1 = n_1, \ldots, X_K = n_K, X_{K+1} = 0, \ldots, X_M = 0 \mid \sum_{i=1}^{K} n_i = H\}$$

we can compute the stationary throughputs of all the stations of $\mathcal{N}'$ when H customers are in the network as

$$\chi_i(H) = \sum_{h=1}^{H} P'_i(h, H) \, \mu_i(h) \tag{7}$$

where $P'_i(h, H)$ is the marginal distribution of the customers in the ith station ($i = 1, \ldots, K$), obtained from the proper summation of the distribution $\pi'_{\boldsymbol{n}}(H)$.

The aggregated throughput of sub-network $\mathcal{K}$ can be written as

$$\chi_{agg}(H) = \sum_{i \in \mathcal{K}} \left[\chi_i(H) \sum_{j \in \mathcal{K}'} r_{i,j} \right] \tag{8}$$

Denoting with s_{eq} a load-dependent station with service intensity equal to $\chi_{agg}(H)$, we can state the following theorem.

Theorem 1. *If station s_{eq} is put in $\mathcal{N}$ in place of the stations that belong to $\mathcal{K}$ in such a way that*
(1) *the routing probabilities from a station $s_i \notin \mathcal{K}$ to s_{eq} are equal to the sum of the routing probabilities from s_i to each station in $\mathcal{K}$, i.e.*

$$r_{i,eq} = \sum_{l \in \mathcal{K}} r_{i,l} \tag{9}$$

(2) *the routing probabilities from s_{eq} to a station $s_j \notin \mathcal{K}$ are equal to the weighted sum of the routing probabilities from each station in $\mathcal{K}$ to s_j,*

$$r_{eq,j} = \frac{\sum_{i \in \mathcal{K}} V_i \, r_{i,j}}{\sum_{l \in \mathcal{K}'} \sum_{h \in \mathcal{K}} V_h \, r_{h,l}} \tag{10}$$

then the resulting new network (denoted by $\mathcal{N}_{eq}$) has measures of interest equal to those of $\mathcal{N}$ when time approaches infinity.

The result expressed by Theorem 1 is well-known for what concerns the characterization of the equivalent server [4,5,11,15]. The proof usually focuses on showing that the normalization constant of the original network $\mathcal{N}$ is identical to that of the reduced network $\mathcal{N}_{eq}$.

Little is instead available in literature for what concerns the modifications that need to be introduced in the routing matrix of network $\mathcal{N}$ in order to provide a detailed and precise specification of network $\mathcal{N}_{eq}$ as defined by (9) and (10) included in the statement Theorem 1. This is due to the fact that these details are not relevant for the computation of the stationary distribution of the whole network; they are instead needed for the transient analysis based on flow equivalent servers that will be discussed later on in this paper. To check the validity of the expressions represented by (9) and (10), it is sufficient to show that the visit ratios computed for the stations that belong to $\mathcal{K}'$ remain identical when considered both within the original network $\mathcal{N}$ and within the reduced network $\mathcal{N}_{eq}$.

Indeed, if we detail the expression represented by (4), and we perform a simple manipulation we obtain

$$\begin{cases} V_i = \sum_{j \in \mathcal{K}'} V_j \, r_{j,i} + \sum_{h \in \mathcal{K}} V_h \, r_{h,i} & i \in \mathcal{K}' \\ \sum_{l \in \mathcal{K}} V_l = \sum_{j \in \mathcal{K}'} V_j \sum_{l \in \mathcal{K}} r_{j,l} + \sum_{h \in \mathcal{K}} V_h \left[1 - \sum_{j \in \mathcal{K}'} r_{h,j} \right] \end{cases} \tag{11}$$

Reorganizing the previous representation and defining (see also [4])

$$V_{eq} = \sum_{j \in \mathcal{K}'} \sum_{h \in \mathcal{K}} V_h \, r_{h,j} \tag{12}$$

we can transform the previous system of equations as follows

$$\begin{cases} V_i = \sum_{j \in \mathcal{K'}} V_j \, r_{j,i} + V_{eq} \frac{\sum_{h \in \mathcal{K}} V_h \, r_{h,i}}{\sum_{j \in \mathcal{K'}} \sum_{h \in \mathcal{K}} V_h \, r_{h,j}} & i \in \mathcal{K'} \\ V_{eq} = \sum_{j \in \mathcal{K'}} V_j \sum_{l \in \mathcal{K}} r_{j,l} \end{cases} \qquad (13)$$

which proves our theorem when we define

$$\begin{cases} r_{eq,eq} = 0 \\ r_{i,eq} = \sum_{l \in \mathcal{K}} r_{i,l} & i \in \mathcal{K'} \\ r_{eq,j} = \frac{\sum_{h \in \mathcal{K}} V_h \, r_{h,j}}{\sum_{l \in \mathcal{K'}} \sum_{h \in \mathcal{K}} V_h \, r_{h,l}} & j \in \mathcal{K'} \end{cases} \qquad (14)$$

4 Use of Flow-Equivalence in Transient Analysis

In the previous section we showed that the key point for applying flow-equivalence is the computation of the service rates $\chi_{agg}(h)$, $1 \le h \le N$, that are required to define the equivalent station. Since the stationary probabilities of a PFQN depend only on the mean service times of the stations, these service rates fully characterize an equivalent server which can take the place of an arbitrary number of aggregated stations without affecting the stationary measures of the system [4,5,11,15].

Unfortunately, the application of the flow equivalence in transient analysis is more difficult and no general results are known for this purpose since the computation of the service rates of a "transient" equivalent server requires the knowledge of the transient distribution of the whole network. In order to clarify this concept, in Section 4.1 we will derive general expressions for aggregation in a transient context and in Section 4.2 we introduce the proposed approximate flow equivalence method.

4.1 Exact Aggregation in Transient Analysis

As the network of queues forms a CTMC, $\boldsymbol{\pi}(t)$ satisfies the well-known ordinary differential equation

$$\frac{d\boldsymbol{\pi}(t)}{dt} = \boldsymbol{\pi}(t) \, \boldsymbol{Q} \qquad (15)$$

where $\boldsymbol{Q}$, the infinitesimal generator, is a square matrix of size $|\mathcal{S}(N,M)| \times |\mathcal{S}(N,M)|$ with the non-null entries defined as

$$q_{n,m} = \begin{cases} \mu_i(n_i) \, r_{i,j} & \boldsymbol{m} \ne \boldsymbol{n}, \boldsymbol{m} = \{n_1, \ldots, n_i - 1, \ldots n_j + 1, \ldots, n_M\} \\ -\sum_{\forall k \ne n} q_{n,k} & \boldsymbol{m} = \boldsymbol{n} \end{cases} \qquad (16)$$

Using the notation introduced in the previous section, let us assume that network $\mathcal{N}$ is split into two sub-networks, one comprising the first $M-1$ stations, that we call $\mathcal{K}$ (stations with indices from 1 up to $M-1$), and the other consisting of the last station only (station with index M). A general state of this network ($\boldsymbol{n}$) can be denoted by means of a pair $\boldsymbol{n} = (\boldsymbol{n'}, n_M)$ so that the entire state space of the network can now be seen as the union of $N+1$ subsets $\mathcal{S}'(H,N,M), H = 0, ..., N$, defined in the following manner

$$\mathcal{S}'(H,N,M) = \left\{ \boldsymbol{n} = (n_1, ..., n_M) \mid n_i \geq 0; \sum_{i=1}^{M-1} n_i = H, \; n_M = N - H; \right\} \quad (17)$$

Let us also assume that the states $\boldsymbol{n}$ of $\mathcal{N}$ are organized in a lexicographical order, so that first we have the state with $n_M = N$, followed by the states with $n_M = (N-1)$ up to the group of states characterized by $n_M = 0$. According to this organization of the state space, the system of differential equations given in (15) can now be divided into $N+1$ sub-systems whose left-hand-sides are characterized by the derivatives of the transient probabilities of the states of the corresponding groups. Let us denote by

$$\tilde{\pi}_H(t) = \sum_{\boldsymbol{n}' \in \mathcal{S}'(H,N,M)} \pi_{(\boldsymbol{n}',N-H)}(t) \quad (18)$$

the probability that there are $H, 0 \leq H \leq N$, clients in $\mathcal{K}$. Then by proper summations of the equations in (15) we can write a system of ordinary differential equations for the quantities defined in (18) in the form of

$$\frac{d\tilde{\boldsymbol{\pi}}(t)}{dt} = \tilde{\boldsymbol{\pi}}(t)\,\tilde{\boldsymbol{Q}}(t) \quad (19)$$

where $\tilde{\boldsymbol{Q}}(t)$ is an $(N+1) \times (N+1)$ matrix, whose non-null entries are given as

$$\tilde{q}_{h,k}(t) = \begin{cases} \chi_{agg}(h,t) & k = h - 1 \\ \mu_M(N-h)(1 - r_{M,M}) & k = h + 1 \\ -\sum_{\forall l \neq h} \tilde{q}_{h,l} & h = k. \end{cases} \quad (20)$$

in which $\chi_{agg}(h,t)$ is the aggregated service rate of the stations in $\mathcal{K}$ at time t if there are h clients in $\mathcal{K}$ and (obviously) $\mu_M(0) = 0$. The term $\chi_{agg}(h,t)$ can be computed according to the following derivation.

Let $\nu_h^{[\boldsymbol{n}']}(t)$ be the conditional probability of finding the sub-network $\mathcal{K}$ in state $\boldsymbol{n}'$, given that there are h customers in it:

$$\nu_h^{[\boldsymbol{n}']}(t) = \frac{\pi_{(\boldsymbol{n}',N-h)}(t)}{\sum_{\boldsymbol{n} \in \mathcal{S}(N,M):n_M=N-h} \pi_{\boldsymbol{n}}(t)}$$

Given a specific state $\boldsymbol{n}'$ of sub-network $\mathcal{K}$, the rate at which customers flow from subnetwork $\mathcal{K}$ to station M is expressed as

$$Y_{\boldsymbol{n}'} = \sum_{l=1}^{M-1} \mu_l(n_l')\, r_{l,M} \quad (21)$$

so that

$$\chi_{agg}(h,t) = \sum_{\boldsymbol{n}' \in \mathcal{S}'(h,N,M)} Y_{\boldsymbol{n}'}\, \nu_h^{[\boldsymbol{n}']}(t) \quad (22)$$

When started from identical initial conditions, the aggregated system (given in (19)) and the original one (given in (15)) lead to the same transient behavior for what concerns the number of clients in $\mathcal{K}$ and with respect to station M.

The derivation of the service rate of the flow equivalent server, that we have proposed for the case of a single sub-network to avoid unneeded complexity, can be easily generalized to the case of any number of sub-networks without changing the essence of the result. The following example illustrates the exact transient aggregation described above.

Example 1. Consider a network of three queues with routing probabilities $r_{1,3} = r_{3,2} = 1, r_{2,1} = r_{2,3} = 1/2$, service rates $\mu_1 = 4, \mu_2 = 3, \mu_3 = 2$ and with two clients. Let us assume that the states are ordered as $|0,0,2|$, $|0,1,1|$, $|1,0,1|$, $|0,2,0|$, $|1,1,0|$, $|2,0,0|$ and that both clients are at the third queue initially. Moreover, denote the transient probabilities of the network by $\pi_i(t)$ where i is one of the states of the network. The system of ordinary differential equations for this model is

$$\frac{d\pi_{|0,0,2|}(t)}{dt} = -\pi_{|0,0,2|}(t)\mu_3 + \pi_{|0,1,1|}(t)\mu_2 r_{2,3} + \pi_{|1,0,1|}(t)\,\mu_1$$

$$\frac{d\pi_{|0,1,1|}(t)}{dt} = \pi_{|0,0,2|}(t)\,\mu_3 - \pi_{|0,1,1|}(t)\,(\mu_2+\mu_3) + \pi_{|0,2,0|}(t)\,\mu_2 r_{2,3} + \pi_{|1,1,0|}(t)\,\mu_1$$
$$\frac{d\pi_{|1,0,1|}(t)}{dt} = \pi_{|0,1,1|}(t)\,\mu_2 r_{2,1} - \pi_{|1,0,1|}(t)\,(\mu_1+\mu_3) + \pi_{|1,1,0|}(t)\,\mu_2 r_{2,3} + \pi_{|2,0,0|}(t)\,\mu_1$$

$$\frac{d\pi_{|0,2,0|}(t)}{dt} = \pi_{|0,1,1|}(t)\,\mu_3 - \pi_{|0,2,0|}(t)\,\mu_2$$
$$\frac{d\pi_{|1,1,0|}(t)}{dt} = \pi_{|1,0,1|}(t)\,\mu_3 + \pi_{|0,2,0|}(t)\,\mu_2 r_{2,1} - \pi_{|1,1,0|}(t)\,(\mu_1+\mu_2)$$
$$\frac{d\pi_{|2,0,0|}(t)}{dt} = \pi_{|1,1,0|}(t)\,\mu_2 r_{2,1} - \pi_{|2,0,0|}(t)\,\mu_1$$

where we highlighted the groups of equations corresponding to specific values of the number of customers at the third station (the first equation refers to the state with all the customers on the third station; the second group of equations corresponds to 1 customer at the third station; and finally the last group to the case when there are no customers at the third station). Summing up the equations for each group we get

$$\frac{d\pi_{|0,0,2|}(t)}{dt} = -\pi_{|0,0,2|}(t)\,\mu_3 + \pi_{|0,1,1|}(t)\,\mu_2 r_{2,3} + \pi_{|1,0,1|}(t)\,\mu_1$$

$$\frac{d[\pi_{|0,1,1|}(t)+\pi_{|1,0,1|}(t)]}{dt} = \pi_{|0,0,2|}(t)\,\mu_3 - [\pi_{|0,1,1|}(t) + \pi_{|1,0,1|}(t)]\,\mu_3$$
$$-[\pi_{|0,1,1|}(t)\mu_2 r_{2,3} + \pi_{|1,0,1|}(t)\mu_1]$$
$$+[\pi_{|0,2,0|}(t)\,\mu_2 r_{2,3} + \pi_{|1,1,0|}(t)\,(\mu_1 + \mu_2 r_{2,3}) + \pi_{|2,0,0|}(t)\,\mu_1]$$

$$\frac{d[\pi_{|0,2,0|}(t)+\pi_{|1,1,0|}(t)+\pi_{|2,0,0|}(t)]}{dt} = [\pi_{|0,1,1|}(t) + \pi_{|1,0,1|}(t)]\,\mu_3$$
$$-[\pi_{|0,2,0|}(t)\,\mu_2 r_{2,3} + \pi_{|1,1,0|}(t)\,(\mu_1 + \mu_2 r_{2,3}) + \pi_{|2,0,0|}(t)\,\mu_1]$$

The left hand sides of these three equations express the derivatives of the probabilities of the aggregated states $|0,2|, |1,1|, |2,0|$ which correspond to lumping together stations 1 and 2. Looking at the right hand sides, we can identify the speed at which the aggregated stations send clients to the third one. Indeed, when the state of the aggregated network is $|1,1|$ the aggregated stations send client to the third queue with intensity $\pi_{|1,0,1|}(t)\mu_1 + \pi_{|0,1,1|}(t)\mu_2 r_{2,3}$

By defining the probability distribution of the aggregated network as

$$\tilde{\pi}_{|0,2|}(t) = \pi_{|0,0,2|}(t) \qquad\qquad \tilde{\pi}_{|1,1|}(t) = \pi_{|0,1,1|}(t) + \pi_{|1,0,1|}(t)$$
$$\tilde{\pi}_{|2,0|}(t) = \pi_{|0,2,0|}(t) + \pi_{|1,1,0|}(t) + \pi_{|2,0,0|}(t)$$

and the conditional probabilities of finding the aggregated stations in a specific state, given the total number of customers in the aggregation as

$$\nu_0^{[|0,0|]}(t) = \frac{\pi_{|0,0,2|}(t)}{\tilde{\pi}_{|0,2|}(t)} \qquad \nu_1^{[|0,1|]}(t) = \frac{\pi_{|0,1,1|}(t)}{\tilde{\pi}_{|1,1|}(t)} \qquad \nu_1^{[|1,0|]}(t) = \frac{\pi_{|1,0,1|}(t)}{\tilde{\pi}_{|1,1|}(t)}$$
$$\nu_2^{[|0,2|]}(t) = \frac{\pi_{|0,2,0|}(t)}{\tilde{\pi}_{|2,0|}(t)} \qquad \nu_2^{[|1,1|]}(t) = \frac{\pi_{|1,1,0|}(t)}{\tilde{\pi}_{|2,0|}(t)} \qquad \nu_2^{[|2,0|]}(t) = \frac{\pi_{|2,0,0|}(t)}{\tilde{\pi}_{|2,0|}(t)}$$

it is possible to re-write the reduced system of differential equations in the following manner

$$\frac{d\tilde{\pi}_{|0,2|}(t)}{dt} = \tilde{\pi}_{|0,2|}(t)\,\tilde{q}_{|0,2|,|0,2|} + \tilde{\pi}_{|1,1|}(t)\,\tilde{q}_{|1,2|,|0,2|}$$

$$\frac{d\tilde{\pi}_{|1,1|}(t)}{dt} = \tilde{\pi}_{|0,2|}(t)\,\tilde{q}_{|0,2|,|1,1|} + \tilde{\pi}_{|1,1|}(t)\,\tilde{q}_{|1,1|,|1,1|} + \tilde{\pi}_{|2,0|}(t)\,\tilde{q}_{|2,0|,|1,1|}$$

$$\frac{d\tilde{\pi}_{|2,0|}(t)}{dt} = \tilde{\pi}_{|1,1|}(t)\,\tilde{q}_{|1,1|,|2,0|} + \tilde{\pi}_{|2,0|}(t)\,\tilde{q}_{|2,0|,|2,0|}$$

where, for example, the rate of the inhomogeneous Markov chain from state $|0,2|$ to state $|1,1|$ is $\tilde{q}_{|0,2|,|1,1|}(t) = \mu_3$, that from state $|2,0|$ to state $|1,1|$ is $\tilde{q}_{|2,0|,|1,1|}(t) = (\nu_2^{[|0,2|]}(t)\,\mu_2 r_{2,3} + \nu_2^{[|1,1|]}(t)\,(\mu_1 + \mu_2 r_{2,3}) + \nu_2^{[|2,0|]}(t)\,\mu_1)$, and that from state $|1,1|$ to itself is $\tilde{q}_{|1,1|,|1,1|}(t) = -(\mu_3 + \nu_1^{[|0,1|]}(t)\,\mu_2 r_{2,3} + \nu_1^{[|1,0|]}(t)\,\mu_1)$ which corresponds to specific instances of (22).

The above 3 differential equations give the exact characterization of the behavior of the original model; the obvious downside of this solution is that we used the transient probabilities of the original network to construct the aggregated network and that the equivalent time-dependent service rates depend on the initial conditions.

It is clear from the previous example and from the preceding more general discussion that exact flow equivalence characterization is difficult to obtain. In particular, we have shown that

- in order to capture the transient behavior of the original network, the aggregated CTMC is time-inhomogeneous,
- the computation of the time-dependent rates of the aggregated CTMC requires the solution of the original model,
- the time-dependent rates depend on the initial state of the original model which precludes the possibility to transport the characterization of the aggregated servers from one experiment to another.

These observations emphasize the limited practical applicability of the above results and the problems of using the flow equivalent approach in transient analysis. Still they provide strong motivation for investigating the possibility of computing approximate solutions using the heuristics that we introduce in the next section.

4.2   Approximate Aggregation for Transient Analysis

Here, we present the simplest strategy that maintains the advantages of the original technique, namely, that the aggregate server is characterized on the basis of the analysis of the aggregated stations in isolation, and that can be used to approximate the transient probabilities of the original model with reduced computational cost.

The idea is to impose $\chi_{agg}(h, t) = \chi_{agg}(h)$, $1 \leq h \leq N$, for every t, i.e., to use information gained from the steady state analysis of the "short-circuited" servers to define the rates of the aggregate station. Since the rates $\chi_{agg}(h)$ can be evaluated using computationally efficient algorithms [9], their cost is negligible with respect to that of the transient analysis. Moreover, this idea guarantees that the approximate transient analysis tends to the correct steady state result.

On the other hand, this strategy corresponds to assume that the rate of service of the equivalent server is not affected by the transient probabilities of the remaining stations of the system and by the position of the customers within the aggregated stations. Indeed, this is a strong assumption that, in general, does not hold but that can be used under certain conditions that we will describe briefly in Section 5.

We briefly mention here that an approach based on information gained from the transient analysis of the "short-circuited" servers is also worth of being investigated. This would lead however to a situation in which the behavior of the whole network is more difficult to define because subsequent periods with different number of customers in the aggregate would need to be synchronized. Hence the overall model would become non-Markovian.

5   Numerical Illustration on a General Network

We tested the method on many networks in which the parameters and the aggregations were selected randomly except for two stations that were used as *observation* points to evaluate the accuracy of the approximation under different initial conditions. All these experiments showed that the more the initial condition of the analysis is such to quickly fill the waiting room of at least one of the aggregated stations the more our method provides inaccurate results.

For every experiment, we compared the transient probabilities of the original and the aggregated models. The results of these comparisons are illustrated by figures with the mean and the variance of relevant quantities at different points in time. Specifically, we chose to focus on the behavior of the number of customers at the non-aggregated stations (the observation points mentioned before). The analysis of the aggregated model has been done analytically for all the experiments, whereas the results for the original network have been obtained via simulation.

We ran a large number of experiments that we cannot report here in detail due to space reasons. Instead we decided to restrict our discussion on the behavior of a single model where, by changing only a few parameter values, it is

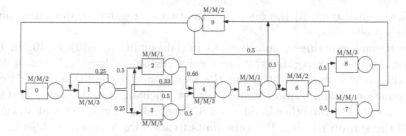

Fig. 1. Random network model

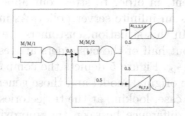

Fig. 2. Aggregated network derived using aggregate servers $s_{0,1,2,3,4}$ and $s_{6,7,8}$

Table 1. Number of servers and service rates of the stations of the random model

station id	0	1	2	3	4	5	6	7	8	9
num. servers	2	3	1	5	3	1	2	1	3	2
service rate	3.5368	4.4287	3.006	3.008	0.8146	0.8504	1.0648	2.3953	0.4324	0.1002

possible to construct scenarios that can be either favorable for suggesting the use of the flow-equivalent servers or adverse to the adoption of our method.

We analyze the system depicted in Figure 1 with the parameters reported in Table 1; we highlighted stations 5 and 9 because they were not considered for aggregation. The total number of costumers was set to 40. Two aggregations are considered: the first, referred as $s_{6,7,8}$, aggregates stations 6, 7 and 8 whereas the second stations 0, 1, 2, 3, and 4 into a single station called $s_{0,1,2,3,4}$. The resulting aggregated model is depicted in Figure 2.

The first experiment places all the customers at station 5 at the beginning of the analysis, i.e., $n_5(0) = 40$. According to this initial condition, on average a half of the customers goes directly to one of the aggregated stations. Observing Figure 3 which depicts the expected numbers of customers at the two stations, it is possible to note that they have a symmetric behavior: while the queue of 5 is getting empty the waiting room of station 9 gets full. The variability of the two phenomena are quite similar since both of them are characterized by a peak around time 40. However, due to the interactions with the other stations the variance of the number of customers at station 9 stabilizes around 2 whereas the same measure for station 5 is almost zero in condition of stability. It is evident that: i) the trajectories generated by the aggregated model are indistinguishable from the original ones; ii) the queues of the two aggregated stations are empty with high probability.

The second experiment considers as initial condition $n_9(0) = 40$. In this situation, both the aggregated stations receive customers directly. Figure 4 illustrates the results obtained by using the second initial condition. In this case the population within station 9 stays almost unchanged and reaches a stable condition soon. On the other hand, even if the number of customers at station 5 is small the station requires 100 time units to stabilize completely. Also in this case the two aggregate servers provide an accurate approximation of the original trajectories.

In our third experiment, in order to stress our approximation method, we assume that station 5 has an infinite server policy. As initial condition all costumers are in station 5. In this situation, customers are placed again far from the aggregated stations but half of the customers arrives, on average, fast into the aggregated station $s_{6,7,8}$. Figure 5 depicts the comparison between the trajectories generated by the original model and those generated by using the two aggregate servers. In this case, looking at the trajectories of station 9, it is possible to see a significant difference between the approximated curves and the original ones. Still, the aggregated model provides a satisfactory picture of the whole phenomenon. In particular, the approximated mean is able to reproduce the slope of the original trajectory even if it overestimates its value. The variance instead is overestimated slightly at the beginning and then underestimated for several time units. However, also in this case the shape of the original curve is reproduced by the approximation.

As a last experiment, we considered again $n_9(0) = 40$ as initial condition and we assumed that station 9 has an infinite server policy whereas station 5 is single server. Figure 6 depicts the results by showing that approximated curves corresponding to the mean and the variance of the number of customers at station 5 fail to reproduce the peak that characterizes its transient.

The experiments show that the flow equivalence approximation works fine when the overall speed of an aggregate (of the flow-equivalent server) is much higher than that of the non-aggregated stations. Opposite is the situation in case of very slow flow equivalent servers which yield poor approximations.

Obviously the really interesting situation is that represented by intermediate configurations for which it would be nice to have criteria to decide whether the transient analysis of the aggregated model is reliable or not. The initial condition for the transient analysis plays an important role in deciding whether certain stations are candidate to be aggregated or not. For sure, one can recommend the aggregation not to include stations which are populated in the initial condition of the model. Assuming that a subnetwork that we want to aggregate is empty in the initial configuration, the speed at which customers can reach the aggregate in the first moments of the transient evolution of the model is critical. Another crucial point is the routing within the aggregation since, by construction, the flow equivalence approximation is not able to represent the variability introduced by the presence of paths involving intermediate stations with different speeds. Thus,

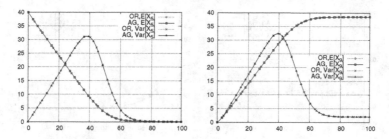

Fig. 3. Original and approximated means and variances of the numbers of customers in station 5 (left) and station 9 (right) computed starting from $X_5 = 40$ and all the other queues empty

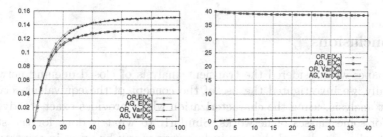

Fig. 4. Original and approximated means and variances of the numbers of customers in station 5 (left) and station 9 (right) computed by starting with $X_9 = 40$ and all the other queues empty

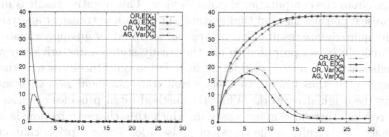

Fig. 5. Original and approximated means and variances of the numbers of customers within station 5 (left) and station 9 (right) computed by considering station 5 as an infinite server and starting with $X_5 = 40$

subnetworks in which the path of a customer is independent both of the station the customer comes from and of the station to which the customer proceeds when it leaves the aggregate are good candidate for aggregation. The aggregation is thus not recommended when the customers quickly accumulate in the aggregates (initially empty) at the beginning of the transient period of interest and when when paths with substantially different speeds are possible within the aggregates. The aggregation is instead reliable when customers are initially far from the aggregated stations and take a substantial time to reach them.

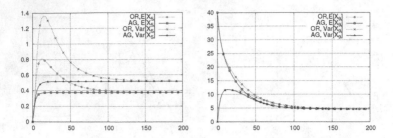

Fig. 6. Original and approximated means and variances of the numbers of customers within station 5 (left) and station 9 (right) computed by considering station 9 as an infinite server and starting with $X_9 = 40$

6 Conclusions

In this paper we considered the transient analysis of closed queuing networks. Specifically, we investigated the use of the concept of the equivalent server in transient analysis, using the characterization which provides exact steady state results for some classes of networks. Similarly to what is done for the steady state behavior, we showed that equivalent servers can be exact in the transient phase as well, but their characterizations require the knowledge of the solution of the whole original network and depend also on the initial location of the customers in the network. Consequently, the exact characterization does not lead to advantages from the computational point of view. This negative result is important because it clarifies the fact that station aggregation in transient analysis can only yield approximate results. For this reason, we opted for an approximate approach. Particularly, we proposed to use the steady state characterization which can be efficiently computed considering in isolation only those stations that we aim to aggregate, thus leading to a significant computational gain with respect analyzing the original network. While the method often provides accurate and reliable results, there are many cases in which this characterization leads to highly inaccurate results in the transient phase. Identifying a priori characteristics of the model that can be exploited to decide whether the the approximation yields reasonable results is an important task that we plan to address in the near future. Moreover, we plan to continue to study this difficult and important problem with the objective of defining a set of criteria that can be tested with limited computational effort to decide on the reliability of the results obtained with flow equivalent approximation suggested in this paper.

Acknowledgments. This work has been supported in part by project "AMALFI - Advanced Methodologies for the Analysis and management of Future Internet" sponsored by Università di Torino and Compagnia di San Paolo, and by project grant Nr. 10-15-1432/HICI from the King Abdulaziz University of the Kingdom of Saudi Arabia.

References

1. Angius, A., Horváth, A.: Product Form Approximation of Transient Probabilities in Stochastic Reaction Networks. ENTCS **277**, 3–14 (2011)
2. Angius, A., Horváth, A.: Approximate transient analysis of queuing networks by decomposition based on time-inhomogeneous markov arrival processes. In: Proc. of 8th International Conference on Performance Evaluation Methodologies and Tools (ValueTools 2014), Bratislava, Slovakia, pp. 1–8, (2014)
3. Angius, A., Horváth, A., Wolf, V.: Approximate transient analysis of queuing networks by quasi product forms. In: Dudin, A., De Turck, K. (eds.) ASMTA 2013. LNCS, vol. 7984, pp. 22–36. Springer, Heidelberg (2013)
4. Balbo, G., Bruell, S.C.: Calculation of the moments of the waiting time distribution of FCFS stations in product form queueing networks. Computer Performance, **4**(2), June 1983
5. Balsamo, S., Iazeolla, G.: An extension of Norton's theorem for queueing networks. IEEE Trans. on Software Eng., SE-8 (1982)
6. Baskett, F., Chandy, K.M., Muntz, R.R., Palacios, F.G.: Open, closed, and mixed networks of queues with different classes of customers. J. ACM **22**(2), 248–260 (1975)
7. Bazan, P., German, R.: Approximate transient analysis of large stochastic models with WinPEPSY-QNS. Computer Networks **53**, 1289–1301 (2009)
8. Boucherie, R.J., Taylor, P.G.: Transient product form distributions in queueing networks. Discrete Event Dynamics Systems: Theory and Applications **3**, 375–396 (1993)
9. Bruell, S.C., Balbo, G.: Computational Algorithms for Closed Queueing Networks. The Computer Science Library, Elsevier North Holland (1980)
10. Buzen, J.P.: Computational algorithms for closed queueing networks with exponential servers. Commun. ACM **16**(9), 527–531 (1973)
11. Chandy, K.M., Herzog, U., Woo, L.: Parametric analysis of queueing networks. IBM Journal of Res. and Dev. **1**(1), 36–42 (1975)
12. Chandy, K.M., Sauer, C.H.: Approximate methods for analyzing queueing network models of computing systems. ACM Comput. Surv. **10**(3), 281–317 (1978)
13. Chandy, K.M., Sauer, C.H.: Computational algorithms for product form queueing networks. Commun. ACM **23**(10), 573–583 (1980)
14. Chen, H., Mandelbaum, A.: Discrete flow networks: Bottleneck analysis and fluid approximations. Mathematics of Operations Research **16**(2), 408–446 (1991)
15. Denning, P.J., Buzen, J.P.: The operational analysis of queueing network models. ACM Comput. Surv. **10**(3), 225–261 (1978)
16. Gordon, W.J., Newell, G.F.: Cyclic queueing networks with restricted length queues. Operations Research **15**(2), 266–277 (1967)
17. Kritzinger, P., van Wyk, S., Krezesinski, A.: A generalization of Norton's theorem for multiclass queueing networks. Perform. Eval. **2**, 98–107 (1982). Elsevier
18. Lavenberg, S.S.: Computer Performance Modeling Handbook. Academic Press, New York (1983)
19. Matis, T.I., Feldman, R.M.: Transient analysis of state-dependent queueing networks via cumulant functions. J. of Applied Probability **38**(4), 841–859 (2001)
20. Whitt, W.: Decomposition approximations for time-dependent Markovian queueing networks. Operations Research Letters **24**, 97–103 (1999)

Model Checking of Open Interval Markov Chains

Souymodip Chakraborty[✉] and Joost-Pieter Katoen

RWTH Aachen University, 52056 Aachen, Germany
souymodip@gmail.com

Abstract. We consider the model checking problem for interval Markov chains with open intervals. Interval Markov chains are generalizations of discrete time Markov chains where the transition probabilities are intervals, instead of constant values. We focus on the case where the intervals are open. At first sight, open intervals present technical challenges, as optimal (min, max) value for reachability may not exist. We show that, as far as model checking (and reachability) is concerned, open intervals does not cause any problem, and with minor modification existing algorithms can be used for model checking interval Markov chains against PCTL formulas.

1 Introduction

Discrete time Markov chains (DTMCs) are useful models for analyzing the reliability and performance of computer systems. A DTMC is defined as a weighted directed graph where the weights on the outgoing transitions define a probability distribution. In general, the precise values of these probabilities may not be always available [9,11,12]. This is precisely the case when transition probabilities are obtained by statistical methods.

Interval Markov chains [9,13] are useful in modeling and verifying probabilistic systems where the value of the transition probabilities are not known precisely. IMCs generalize discrete time Markov chains by allowing intervals of possible probabilities on the state transitions in order to capture the system uncertainty more faithfully. For example, instead of specifying that the probability of moving from state s to t is 0.5, one can specify an interval $[0.3, 0.7]$ which captures the uncertainty in the probability of moving from state s to t. Uncertainty in the model may occur due to various reasons [12]. In some cases, the transition probabilities may depend on an unknown environment, and are approximately known, in other cases the interval may be introduced to make the model more robust.

There are two prevalent semantics of interval Markov chains. *Uncertain Markov Chains* (UMC) [9,11] is an interpretation of interval Markov chains as set of (possibly uncountably many) discrete time Markov chains where each element of the set is a DTMC whose transition probabilities lie within the interval range defined by the IMC. In the other semantics, called *Interval Markov Decision Processes* (IMDP) [11], the uncertainty of the transition probabilities are resolved non-deterministically. It requires the notion of *scheduler*, which chooses

© Springer International Publishing Switzerland 2015
M. Gribaudo et al. (Eds.): ASMTA 2015, LNCS 9081, pp. 30–42, 2015.
DOI: 10.1007/978-3-319-18579-8_3

a distribution, each time a state is visited in an execution, from a (possibly uncountable) set of distributions defined by the intervals on the transitions.

The logic *probabilistic computation tree logic* (PCTL) [8], extends the temporal logic CTL [7] with probabilities. This allows us to express properties like "after a request for a service, there is 99% chance of fulfilling the request". PCTL formulas are interpreted over DTMCs and model checking on DTMCs can be done in PTIME. The problem of model checking PCTL properties for IMCs was studied in [11], it provides a PSPACE algorithms for both UMC and IMDP semantics for interval Markov chains. Furthermore, NP and co-NP hardness was shown for model checking in UMC semantics and PTIME hardness for IMDP semantics which follows from PTIME hardness of model checking PCTL formulas on DTMCs. [4] improved the upper bound and showed that model checking problem for IMDP semantics is in co-NP. This result is shown for a richer class of logic, called ω-*PCTL*, which allow Büchi and co-Büchi properties in the formula.

In the literature, the intervals of IMCs are always assumed to be closed. This assumption is sensible from the model checking perspective in IMDP semantics as models with open interval may not have an optimal value of satisfying a temporal property. The focus of this paper is to study IMDP semantics of IMCs with open intervals. We will later contrast it with the UMC semantics, and will see that the existing algorithm is applicable for IMCs with open intervals, but its outcome may vary with the model at hand. The main intuition is that the value of reachability property in a IMC with open intervals can be made arbitrarily close to the value of the property obtained by *closing* the intervals. We use this observation to show the equivalence between model checking IMCs with open interval and IMCs with closed intervals.

2 Interval Markov Chains

Definition 1. *Let $\mathcal{I}$ be the set of intervals (open or closed) in the range $[0,1]$. The subsets $\mathcal{I}_0 \triangleq \{(a,b] \mid 0 \le a < b \le 1\}$, $\mathcal{I}_1 \triangleq \{(a,b) \mid 0 \le a < b \le 1\}$, $\mathcal{I}_2 \triangleq \{[a,b) \mid 0 \le a < b \le 1\}$ and $\mathcal{I}_3 \triangleq \{[a,b] \mid 0 \le a \le b \le 1\}$. $\mathcal{I} = \bigcup_{i \in \{0,1,2,3\}} \mathcal{I}_i$.*

Let $I \triangleq \langle a, b \rangle$ be an interval in $\mathcal{I}$, where $\langle \in \{(, [\} \text{ and } \rangle \in \{),]\}$. The lower bound $I \downarrow = a$ and upper bound is $I \uparrow = b$. Point intervals ($[a,a]$) are closed intervals where the upper and lower bounds are equal. The closure of an interval I, denoted by $\bar{I}$, is the smallest closed interval that includes I.

Definition 2. *A discrete time Markov chain (DTMC) is a tuple $M = (S, L, \delta)$ where S is a finite set of states, $L : S \to 2^{AP}$ is a labeling function (AP is the set of atomic propositions), $\delta : S \to S \to [0,1]$ is a transition probability matrix, such that for all $s \in S$, $\sum_{t \in S} \delta(s)(t) = 1$.*

For simplicity of notation we will use the un-Curry notation $\delta(s,t)$ for $\delta(s)(t)$. A path π of a DTMC M is an infinite sequence of states $\pi = s_0 s_1 \ldots$ such that for all $i \ge 0$, $\delta(s_i, s_{i+1}) > 0$. The ith state of the path π is denoted by $\pi_i = s_i$. Let Ω_s be the set of paths starting from state s. The cylinder (open) set $Cyl(\rho)$ is the set of all paths with ρ as prefix. Let $\mathcal{B}$ be the smallest Borel σ-algebra defined on the cylinder sets. Let ρ be a finite sequence of states $s_0 s_1 \ldots s_n$ such that $\delta(s_i, s_{i+1}) > 0$ for all $0 \le i < n$. The unique measure μ is thus induced from δ as, $\mu(Cyl(\rho)) = \delta(s_0, s_1) \cdot \delta(s_1, s_2) \ldots \cdot \delta(s_{n-1}, s_n)$.

Definition 3. *An Interval Markov chain (IMC) is a tuple $\mathcal{M} \triangleq (S, L, \delta)$, where S is a (finite) set of states and L is a labeling function $L : S \to 2^{AP}$, where AP is the set of atomic propositions. δ is a function $\delta : S \to \mathcal{D}$, where $\mathcal{D}$ is the set of functions from the set of states to the set of intervals $\mathcal{I}$, i.e., $\mathcal{D} = S \to \mathcal{I}$.*

As before, we will use the un-Curry notation $\delta(s, t)$ for $\delta(s)(t)$. For a state s, the probability of a single step from s to t lies in the interval $\delta(s, t)$. Thus an IMC defines a collection of Markov chains, where the single step transition probability of moving from state s to t lies in the interval $\delta(s, t)$. Not every IMC defines a collection of Markov chains. Thus, we have the notion of realizability.

Definition 4. *Let $\mathcal{M} = (S, L, \delta)$ be an IMC with states $S = \{s_1, \ldots, s_m\}$. Let $D^{\mathcal{M}}$ be the set of $m \times 1$ vectors $\boldsymbol{d}$, such that $\boldsymbol{d}^T \cdot \boldsymbol{1} = 1$, which represents the set of distributions on states of $\mathcal{M}$. Where $\mathcal{M}$ is fixed we denote the set as D.*

$\mathcal{M}$ is said to be realizable if for each set of intervals defined by $\delta(s)$, there exists a distribution $\boldsymbol{d}$ such that for all $i \in [1, m]$ $\boldsymbol{d}_i$ (the ith component of $\boldsymbol{d}$) is in $\delta(s, s_i)$. The distribution $\boldsymbol{d}$ is said to be a solution of $\delta(s)$. Let $sol(s)$ be the set of solutions of $\delta(s)$.

Next we give two semantics of IMCs: 1) *Uncertain Markov chains* (UMC), 2) *Interval Markov decision process* (IMDP).

Definition 5. (Uncertain Markov chain semantics) *An IMC $\mathcal{M} = (S, L, \delta)$ represents a set of DTMCs, denoted by $[\mathcal{M}]_u$, such that for each DTMC $M = (S, L, \delta_M)$ in $[\mathcal{M}]_u$, $\delta_M(s)$ is a solution of $\delta(s)$ for every state $s \in S$. In UMC semantics, we assume that nature non-deterministically picks a solution of $\delta(s)$ for each state $s \in S$, and then all transitions behave according to the chosen transition probability matrix.*

To define interval Markov decision process semantics, we need the notion of *schedulers*. The schedulers resolve the non-determinism at each state s by choosing a particular distribution from $sol(s)$.

Definition 6. *A scheduler of an IMC $\mathcal{M} = (S, L, \delta)$ is a function $\eta : S^+ \to D^{\mathcal{M}}$, such that for every finite sequence of states $\pi \cdot s$ of $\mathcal{M}$, $\eta(\pi \cdot s)$ is a solution of $\delta(s)$.*

A path $w = s_0 s_1 s_2 \ldots$ of an IMC $\mathcal{M}$ is an infinite sequence of states. A path w starting from a state s (i.e., $w_0 = s$) is said to be *according* to the scheduler η if for all $i \ge 0$, $\eta(w_0, \ldots, w_i)(w_{i+1}) > 0$. A scheduler is *memoryless* if the choice of the distribution depends solely on the current state, that is, $\eta : S \to D^{\mathcal{M}}$.

Definition 7. (Interval Markov decision process semantics) *In IMDP semantics, before every transition from a state s of a IMC $\mathcal{M} = (S, L, \delta)$, nature chooses a solution of $\delta(s)$ and then takes a one-step probabilistic transition*

*according to the chosen distribution. In other words, nature chooses a scheduler
η which then defines a DTMC M. The set of all DTMC in this semantics is
denoted by $[\mathcal{M}]_d$.*

Obviously, for any IMC $\mathcal{M}$ we have:

$$[\mathcal{M}]_u \subseteq [\mathcal{M}]_d.$$

Given an IMC $\mathcal{M}$ and a state s, let σ-algebra $(\Omega_s, \mathcal{F})$ be the smallest σ-algebra
on the cylinder sets of Ω_s, where Ω_s is the set of infinite paths starting from s.
For each scheduler η we have a probability measure Pr^η (also denoted by $\mu_{\mathcal{M}}^\eta$)
on the events in $\mathcal{F}$.

3   Probabilitic Computation Tree Logic

Probabilistic computation tree logic (PCTL) [8] replaces the path quantifiers in
CTL by probabilistic operators. It has the following syntax:

$$f ::= a \mid \sim f \mid f \wedge f \mid \mathsf{P}_{\bowtie p} g$$
$$g ::= \mathsf{X} f \mid f \ \mathsf{U} \ f$$

where $a \in AP$, f is called a state formula, g is called a path formula, $\bowtie \in \{<,$
$\leq, >, \geq\}$ and p is a rational number in $[0, 1]$. The PCTL semantics is define on
DTMCs. A DTMC M satisfies a state formula f at a state s if:

$$
\begin{aligned}
M, s &\models a & &\text{iff } a \in L(s) \\
M, s &\models \sim f & &\text{iff } M, s \not\models f \\
M, s &\models f_1 \wedge f_2 & &\text{iff } M, s \models f_1 \text{ and } M, s \models f_2 \\
M, s &\models \mathsf{P}_{\bowtie p} g & &\text{iff } Pr\{s \models g\} \bowtie p,
\end{aligned}
$$

where $\{s \models g\} = \{w \mid w_0 = s \text{ and } M, w \models g\}$. A path formula g is true for a
path w of M if:

$$
\begin{aligned}
M, w &\models \mathsf{X} f & &\text{iff } M, w_1 \models f \\
M, w &\models f_1 \ \mathsf{U} \ f_2 & &\text{iff } \exists i : M, w_i \models f_2 \text{ and } \forall j < i : M, w_j \models f_1
\end{aligned}
$$

We will denote the satisfaction relation by $s \models f$ (and $w \models g$) when M is
fixed. Next we define the satisfaction relation of a PCTL formula f for an IMC
$\mathcal{M}$ for the two semantics. In UMC semantics, $\mathcal{M}, s \models_u f$ iff for every DTMC
$M \in [\mathcal{M}]_u$, $M, s \models f$. Note that for a PCTL formula f, $\mathcal{M}, s \models_u f$ does not
imply $\mathcal{M}, s \not\models_u \sim f$. In IMDP semantics, the satisfaction of a PCTL formula f
by a state s of $\mathcal{M}$ ($\mathcal{M}, s \models_d f$) is the same as for a DTMC except the formula
with probabilistic operator, which is as follows:

$$\mathcal{M}, s \models \mathsf{P}_{\bowtie p} g \text{ iff } \forall \eta : Pr_{\mathcal{M}}^\eta \{w \mid w_0 = s \text{ and } M, w \models g\} \bowtie p$$

Particularly,

$$
\begin{aligned}
s &\models \mathrm{Pr}_{\leq c}\, g \text{ iff } \sup_\eta Pr^\eta(s \models g) \leq c \\
s &\models \mathrm{Pr}_{< c}\, g \text{ iff } \sup_\eta Pr^\eta(s \models g) < c \\
s &\models \mathrm{Pr}_{\geq c}\, g \text{ iff } \inf_\eta Pr^\eta(s \models g) \geq c \\
s &\models \mathrm{Pr}_{> c}\, g \text{ iff } \inf_\eta Pr^\eta(s \models g) > c
\end{aligned}
\tag{1}
$$

Thus if event $E \in \mathcal{F}$ defines a set of paths, we are interested in the values

$$\inf_{\eta} Pr^{\eta}_{\mathcal{M}}(E) \quad \text{and} \quad \sup_{\eta} Pr^{\eta}_{\mathcal{M}}(E)$$

Open intervals present a problem for model checking in IMDP semantics. There might not exist a scheduler that gives the optimal values. Consider the reachability problem for IMCs in the following example:

Example 1. It is possible that an optimal scheduler may not exist for IMCs with open intervals. Consider the following example Figure 1, E is the set of paths that eventually reach the state s_1 from s_0. $\inf_{\eta} Pr^{\eta}(E) = 0.6$, but no scheduler gives the probability of reaching s_1 from s_0 as 0.6. The reason for this is the open lower bound of $(0.3, 1]$.

4 ϵ-Approximate Scheduler for Reachability

In this section we consider the reachability problem in IMDP semantics for IMCs with open intervals. As observed in the previous example, an optimal scheduler may not exists, thus we will construct ϵ-approximate schedulers.

An IMC is called a *closed* IMC if the probability interval of every transition is closed. We can obtain a closed IMC from an arbitrary IMC by taking the closure of the probability intervals.

Definition 8. *Given an IMDP* $\mathcal{M} \triangleq (S, L, \delta)$, *a closed IMDP* $\bar{\mathcal{M}}$ *is defined as* (S, L, δ'), *where for every* s, t, $\delta'(s, t) = \bar{\delta}(s, t)$.

Example 2. The closed IMC $\bar{\mathcal{M}}$ for $\mathcal{M}$ in the example 1 is shown below:

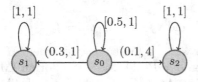

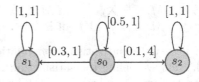

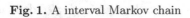

Fig. 1. A interval Markov chain   **Fig. 2.** A closed interval Markov chain

Evidently, if an IMC $\mathcal{M}$ is realizable then $\bar{\mathcal{M}}$ is also realizable.

Definition 9. *Basic feasible solution (BFS). Given a set of closed intervals* $R \triangleq \{I_1, \ldots, I_m\}$ *a basic feasible solution* $\boldsymbol{d}$ *is an* $m \times 1$ *vector, such that there exists a set* $H \subseteq R$ *with* $|H| \geq |R| - 1$ *and for all* $I_i \in H$, $\boldsymbol{d}_i = I_i\!\downarrow$ *or* $\boldsymbol{d}_i = I_i\!\uparrow$, *and* $\boldsymbol{d}^T \cdot \boldsymbol{1} = 1$.

BFSs of a set of intervals $\mathcal{J}$ that contains open intervals are the BFSs of the set of closed intervals $\bar{\mathcal{J}}$. We have the following observation.

Proposition 1. *Every solution of a set of (open or closed) intervals, can be represented as the convex combination of the BFSs.*

Proposition 2 ([4]). *Let $\mathcal{M}$ be a closed IMC, and E be an event defining the reachability of some set of states $T \subseteq S$. There exists a memoryless scheduler η such that the probability of the event E is optimal.*

The proposition says that, if $\mathcal{M}$ is closed then we have a scheduler $\eta : S \to D^{\mathcal{M}}$ such that $Pr^{\eta}(E) = \inf_{\eta'} Pr^{\eta'}(E)$ (or $\sup_{\eta'} Pr^{\eta'}(E)$), and η chooses at each state s one of the BFSs of $\delta(s)$ (pure scheduler). The proposition follows directly from the existence of an optimal scheduler for reachability in Markov Decision Processes [2].

The main theorem of this paper is as follows:

Theorem 1. *Let E be the event describing the set of paths of an IMC $\mathcal{M}$ starting from a state s and eventually reaching some goal states T. Then:*

$$\forall \varepsilon > 0 \;\; \exists \hat{\eta} : |\min_{\eta} Pr^{\eta}_{\mathcal{M}}(E) - Pr^{\hat{\eta}}_{\mathcal{M}}(E)| \le \varepsilon$$

and

$$\forall \varepsilon > 0 \;\; \exists \hat{\eta} : |\max_{\eta} Pr^{\eta}_{\mathcal{M}}(E) - Pr^{\hat{\eta}}_{\mathcal{M}}(E)| \le \varepsilon$$

Proof. Let $\mathcal{M} \triangleq (S, L, \delta)$ and $\bar{\mathcal{M}} \triangleq (S, L, \delta')$. $\bar{\mathcal{M}}$ is closed, thus by Prop. 2 an optimal scheduler exists. Let $\overset{*}{\eta}$ be an optimal scheduler that minimizes $Pr^{\eta}_{\bar{\mathcal{M}}}(E)$. Furthermore, $\overset{*}{\eta}$ is memoryless, deterministic and chooses one of the BFS of $\delta'(s)$ at each state s. Hence, $\overset{*}{\eta}$ induces a DTMC on $\bar{\mathcal{M}}$, and $\overset{*}{\eta}(s, t)$ defines the single step transition probability from a state s to a state t.

Let the stochastic matrix $\overset{*}{P}$ be such that each row is identified with a state of $\bar{\mathcal{M}}$. We have :

$$\overset{*}{P}(s, t) = \overset{*}{\eta}(s, t) \quad \text{if } s \notin T \quad \text{and} \quad \overset{*}{P}(s, s) = 1 \quad \text{if } s \in T \tag{2}$$

Let $A = (1 + \overset{*}{P} + (\overset{*}{P})^2 + (\overset{*}{P})^3 \ldots)$, A is well-defined stochastic matrix as the series converges. Let $\gamma = \|A\|_{\infty}$.

Now we are in a position to define a scheduler $\hat{\eta}$ for the IMC $\mathcal{M}$. The scheduler $\hat{\eta}$ is a function, $\hat{\eta} : S \times \mathbb{N} \to D^{\mathcal{M}}$. We assume that there are no positive point intervals. (We can set the value of $\hat{\eta}$ if point intervals are present.) Define the following:

$$
\begin{aligned}
Q_s &= \{t \mid \overset{*}{\eta}(s, t) > 0, \; \overset{*}{\eta}(s, t) \notin \delta(s)\} \\
L_s &= \{t \mid \overset{*}{\eta}(s, t) \in \delta(s, t), \; \overset{*}{\eta}(s, t) = \delta(s, t)\!\downarrow\} \\
R_s &= \{t \mid \overset{*}{\eta}(s, t) \in \delta(s, t), \; \overset{*}{\eta}(s, t) = \delta(s, t)\!\uparrow\} \\
I_s &= \{t \mid \overset{*}{\eta}(s, t) \in \delta(s, t), \; \overset{*}{\eta}(s, t) \ne \delta(s, t)\!\uparrow, \; \overset{*}{\eta}(s, t) \ne \delta(s, t)\!\downarrow\} \\
\rho &= \min\{\{x \mid \exists s, \exists t \in L_s \cup I_s : x = \overset{*}{\eta}(s, t) - \delta(s, t)\!\downarrow\}, \\
&\qquad \{x \mid \exists s, \exists t \in R_s \cup I_s : x = \delta(s, t)\!\uparrow - \overset{*}{\eta}(s, t)\}, \\
&\qquad \{x \mid \exists s, \exists t \in Q_s : x = \delta(s, t)\!\uparrow = \delta(s, t)\!\downarrow\}\}
\end{aligned}
$$

Observe that ρ is a constant of the model $\mathcal{M}$. Let $\hat{\eta}$ be defined as follows:

– Let $t \in Q_s$. This implies $\overset{*}{\eta}(s,t) = \delta(s,t)\uparrow$ or $\overset{*}{\eta}(s,t) = \delta(s,t)\downarrow$. If $\overset{*}{\eta}(s,t) = \delta(s,t)\uparrow$ then $\delta(s,t)$ is open from above and $\hat{\eta}(s,n,t) = \overset{*}{\eta}(s,t) - 2^{-n}\frac{\kappa\rho}{|Q_s|}$, where $\kappa = \frac{\varepsilon}{1+\gamma}$. Similarly, if $\overset{*}{\eta}(s,t) = \delta(s,t)\downarrow$ then $\delta(s,t)$ is open from below and $\hat{\eta}(s,n,t) = \overset{*}{\eta}(s,t) + 2^{-n}\frac{\kappa\rho}{|Q_s|}$.

– Let $t \in R_s$ and $\alpha \triangleq \sum_{t \in Q_s} \hat{\eta}(s,n,t) - \overset{*}{\eta}(s,t)$. If $\alpha < 0$ then for all $t \in R_s \cup I_s$,

$\hat{\eta}(s,n,t) = \overset{*}{\eta}(s,t) + \frac{\alpha}{|R_s \cup I_s|}$ and for $t \in L_s$, $\hat{\eta}(s,n,t) = \overset{*}{\eta}(s,t)$. If $\alpha > 0$ then for all $t \in L_s \cup I_s$, $\hat{\eta}(s,n,t) = \overset{*}{\eta}(s,t) + \frac{\alpha}{|L_s \cup I_s|}$ and for $t \in R_s$, $\hat{\eta}(s,n,t) = \overset{*}{\eta}(s,t)$. If $\alpha = 0$ then for all $t \in L_s \cup I_s \cup R_s$, $\hat{\eta}(s,n,t) = \overset{*}{\eta}(s,t)$.

It remains to prove that $\boldsymbol{d} = \overset{*}{\eta}(s,n)$, defined above, is a solution to $\delta(s)$. From the construction it follows that $\sum_{t \in S} \boldsymbol{d}_t = 1$ and hence it is a valid distribution on the states of the IMC $\mathcal{M}$. Consider the following cases: $t \in Q_s$ and $\overset{*}{\eta}(s,t) = \delta(s,t)\uparrow$, the upper bound of $\delta(s,t)$ is open. The lower bound of $\delta(s,t)$ is strictly smaller than $2^{-n}\kappa\rho$ for any $n \in \mathbb{N}$ i.e., $\delta(s,t)\downarrow < \kappa\rho$ since ρ is at the most as large as the smallest interval in $\mathcal{M}$. Thus $\boldsymbol{d}_t \in \delta(s,t)$. Similarly, for every $t \in Q_s$, $\boldsymbol{d}_t \in \delta(s,t)$. Suppose $\alpha < 0$, then $R_s \cup I_s$ is not empty, else $\delta(s)$ will not be realizable. The changes to the probability for a transition s to t, where $t \in R_s \cup I_s$ is small enough so that $\boldsymbol{d}_t \in \delta(s,t)$. Thus, for every t, $\boldsymbol{d}_t \in \delta(s,t)$, or equivalently $\boldsymbol{d}$ is a solution to $\delta(s,t)$. Identical argument holds when $\alpha > 0$.

Let $\hat{P}_n$ be a sub-stochastic matrix defined as follows: $\hat{P}_n(s,t) = \hat{\eta}(s,t)$ if $\overset{*}{P}(s,t) > 0$ else $\hat{P}_n(s,t) = 0$. In other words, $\hat{P}_n(s,t) > 0$ if the state t is in $support(\overset{*}{\eta}(s))$.

$$\hat{P}_n = \overset{*}{P} + P_n \tag{3}$$

where $|P_n(s,t)| \le 2^{-n}\kappa\rho$ for every (s,t).

Let $\overset{*}{\eta}$ and $\hat{\eta}$ induce DTMCs M' and M on the IMCs $\bar{\mathcal{M}}$ and $\mathcal{M}$, respectively. Let the corresponding σ-algebra be $\mathcal{S} \triangleq (\Omega_s, \mathcal{F}, \overset{*}{\mu})$ and $\mathcal{S}' \triangleq (\Omega_s, \mathcal{F}, \hat{\mu})$, where s is some state of $\mathcal{M}$ and Ω_s is the set of paths starting from state s. Define $\overset{*}{R} \triangleq \{w \in \Omega_s \mid w$ is according to $\overset{*}{\eta}\}$ and $\hat{R} \triangleq \{w \in \Omega_s \mid w$ is according to $\hat{\eta}\}$, i.e., $\overset{*}{R}$ and R are set of paths in M' and M, respectively. Let $B \in \mathcal{F}$ be the event of reaching the goal states T, and $E = \overset{*}{R} \cap B$ and $E' = \hat{R} \cap B$. It follows from the construction that $E \subseteq E'$. Define $A_i \triangleq \{w \mid \exists u \in E : w_0 \dots w_i = u_0 \dots u_i$ and $\overset{*}{\eta}(w_i, w_{i+1}) = 0, \hat{\eta}(w_i, i, w_{i+1}) > 0\}$. Let $A = \bigcup_i A_i$. It is easy to see that, $E' \cap \bar{A} = E$. We will first show that the event A has a very small probability measure in $\mathcal{S}'$:

$$\hat{\mu}(A) = Pr_M^{\hat{\eta}}(A) = \sum_{i=0} Pr_M^{\hat{\eta}}(A_i)$$

If $w \in A_i$ then $\delta(w_i, w_{i+1})\uparrow > 0$ and $\overset{*}{\eta}(s,t) = 0$. Thus,

$$Pr_M^{\hat{\eta}}(A_i) \le 2^{-i}\kappa\rho \quad \text{or} \quad Pr_M^{\hat{\eta}}(A) \le \kappa\rho$$

Thus,

$$\hat{\mu}(A) \leq \kappa\rho \tag{4}$$

We will now show that the probability of E' can be made infinitesimally close to the probability of E. Formally, we will show, $|\hat{\mu}(E') - \bar{\mu}(E)| \leq \varepsilon$. The left hand side can be written as:

$$\begin{aligned}|\hat{\mu}(E') - \bar{\mu}(E)| &= |\hat{\mu}(E' \cap A) + \hat{\mu}(E' \cup \bar{A}) - \bar{\mu}(E)| \\ &\leq |\hat{\mu}(E) - \bar{\mu}(E)| + \kappa\rho\end{aligned} \tag{5}$$

That is, we restrict to the paths that belong to E. Let x_s^n denote the probability of reaching the goal states T at the nth step in M' from the state s. Let E_n be the event of reaching the goal states T at the nth step in the Markov chain M such that $E_n \subseteq E$ and thus $\bigcup_n E_n = E$. Let $y_s^n = \hat{\mu}(E_n)$. Thus, we can write the following:

$$x_s^{n+1} = \sum_{t \in support(\overset{*}{\eta}(s))} \overset{*}{P}(s,t)x_t^n,$$

$$y_s^{n+1} = \sum_{t \in support(\overset{*}{\eta}(s))} \hat{P}_n(s,t)y_t^n.$$

Or, using vector notation, $\boldsymbol{x}_{n+1} = \overset{*}{P}\boldsymbol{x}_n$ and $\boldsymbol{y}_{n+1} = \hat{P}_n\boldsymbol{y}_n$. Therefore:

$$\begin{aligned}\boldsymbol{y}_{n+1} - \boldsymbol{x}_{n+1} &= \overset{*}{P}(\boldsymbol{y}_n - \boldsymbol{x}_n) + P_n\boldsymbol{y}_n && \text{from equation (3)} \\ &\leq \overset{*}{P}(\boldsymbol{y}_n - \boldsymbol{x}_n) + 2^{-n}\kappa\rho\boldsymbol{1} \\ &\leq 2^{-n}\kappa\rho(1 + \overset{*}{P} + \overset{*}{P}^2 + \ldots)\boldsymbol{1}\end{aligned}$$

Thus, $\|\boldsymbol{y}_{n+1} - \boldsymbol{x}_{n+1}\|_\infty \leq 2^{-n}\kappa\rho\gamma$.

We have,

$$|\hat{\mu}(E) - \bar{\mu}(E)| \leq |\sum_n (y_s^n - x_s^n)| \leq \sum_n 2^{-n}\kappa\rho\gamma \leq \kappa\rho\gamma$$

Combining this with equation (5) we can conclude:

$$|\hat{\mu}(E') - \bar{\mu}(E)| \leq (1+\gamma)\kappa\rho \leq \varepsilon$$

By similar argument we conclude $\forall \varepsilon > 0 \; \exists \hat{\eta} : |\max_\eta Pr_{M'}^\eta(E) - Pr_M^{\hat{\eta}}(E)| \leq \kappa$.

Corollary 1. *Let E be the set of paths that reach some goal states T of IMC $\mathcal{M}$. Then:*

$$\min_\eta Pr_{\bar{\mathcal{M}}}^\eta(E) = \inf_\eta Pr_{\mathcal{M}}^\eta(E) \text{ and } \max_\eta Pr_{\bar{\mathcal{M}}}^\eta(E) = \sup_\eta Pr_{\mathcal{M}}^\eta(E).$$

Proof. We need to show $\forall \kappa > 0 \; \exists \hat{\eta} : |\min_\eta Pr_{M'}^\eta(E) - Pr_M^{\hat{\eta}}(E)| \leq \kappa$. Observe that, $\hat{\eta}$ is also a scheduler of M', thus, $Pr_M^{\hat{\eta}}(E) - \min_\eta Pr_{M'}^\eta(E) \leq \kappa$. Similarly, for all $\kappa > 0$ there exists a scheduler $\hat{\eta}$ of M such that $\max_\eta Pr_{M'}^\eta(E) - Pr_M^{\hat{\eta}}(E) \leq \kappa$.

Example 3. In UMC semantics, the nature picks the probability transition matrix and the model behaves according to it. The infimum (or supremum) probability of reaching some state is different than the infimum probability in IMDP semantics. This becomes apparent in the following IMC with an open interval:

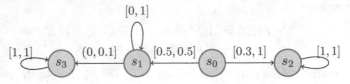

The minimum and maximum probability of reaching state s_3 from s_0 in the UMC semantics is 0.5. But for any $\epsilon > 0$ there exists a scheduler for which the probability of reaching s_3 is smaller than ϵ. That is, the infimum of the probability of reaching state s_3 is 0.

5   PCTL Model Checking

In this section we briefly recall PCTL model checking on DTMC and IMCs with closed intervals (for the two semantics), and then show how to use the result of previous section to do model checking for IMCs with open intervals.

Model checking of PCTL [1,6] formula f on DTMC M proceeds much like the CTL model checking on Kripke structures [5]. The satisfiability of a (state) sub-formula f' of f for a state s of M is iteratively calculated and the labeling functions are updated accordingly. For example, for the until formula $f = P_{\bowtie p}(f_1 \cup f_2)$ and a state s, the formula f is added to the label of s iff the probability of reaching states with label f_2, via states with label f_1 satisfies $\bowtie p$. This can be done in polynomial time by solving linear constraints. Finally, a state $s \models f$ if $f \in L(s)$ and the model checking problem can be solved in polynomial time.

Model checking in UMC semantics uses the *existential* theory of reals [10]. An IMC $\mathcal{M}, s \models_u f$ in UMC semantics iff for all DTMC $M \in [\mathcal{M}]_u$, $M, s \models f$, or equivalently, $\mathcal{M}, s \not\models_u f$ iff there exists a $M \in [\mathcal{M}]_u$ such that $M, s \models \sim f$. Basically, we use parameters to encode the transition probabilities which are constrained by the intervals and construct a formula Γ in existential theory of reals such that Γ is satisfiable iff there exists a $M \in [\mathcal{M}]_u$ such that $M, s \models \sim f$ [4]. Observe, that the presence (or absence) of open intervals does not affect the algorithm and the algorithm operates in PSPACE.

Model checking in IMDP semantics is done by first transforming the IMC into an Markov decision process (MPD) and then doing model checking on the MDP [2]. Let $\mathcal{M} = (S, L, \delta)$ be a *closed* IMC and for each state $s \in S$, let B_s be the set of basic feasible solution of $\delta(s)$. Let $D_{\mathcal{M}} = (S, L, \mu)$ be the MDP with $\mu : S \to S \to [0, 1]$, where $\mu(s) = B_s$. From Proposition 1, we can deduce that, a DTMC $M \in [\mathcal{M}]_d$ iff M is induced by some scheduler η of $D_{\mathcal{M}}$. Model checking of MDP proceeds the same way as model checking of DTMC. We iteratively update the labels of the state with (state) sub-formulas. Conjunctions and disjunctions are handled as in the DTMC model checking. Interesting cases are formulas with probabilistic operator and negations. Let g be a path formula and $\mathsf{P}_{\succ p}g$ (or $\mathsf{P}_{\prec p}g$) is added to the label of a state $s \in S$, iff

$$\min_{\eta} Pr^{\eta}_{D_{\mathcal{M}}}(s \models g) \succ p \text{ (or } \max_{\eta} Pr^{\eta}_{D_{\mathcal{M}}}(s \models g) \prec p)$$

where $\succ \in \{\geq, >\}$ ($\prec \in \{\leq, <\}$). This is done by solving a linear optimization problem. We use the following proposition to handle formulas with negations.

Proposition 3. *For any* $E \in \mathcal{F}$ *of* $(\Omega_s, \mathcal{F})$ *on MDP* M,

$$\inf_{\eta} Pr^{\eta}(E) = 1 - \sup_{\eta} Pr^{\eta}(\bar{E})$$

Thus, model checking MDPs boils down to solving successive reachability optimization problems. Note that direct application of this method to IMCs with open interval is not possible since no scheduler exists which may yields the value $\inf_{\eta} Pr^{\eta}_{D_{\mathcal{M}}}(s \models g)$.

In the rest of the section we use the above mentioned model checking mechanism to show that model checking IMCs with open interval in IMDP semantics, reduces to model checking its closure.

Theorem 2. *Given a PCTL formula* f *and an IMC* $\mathcal{M}$,

$$\mathcal{M}, s \models f \text{ iff } \bar{\mathcal{M}}, s \models f$$

Proof. We assume that $\mathcal{M}$ has open intervals. We proceed by induction on the structure of the formula f. We have the following cases:

1. Let $f := a$. The labeling function of s in $\mathcal{M}$ and $\bar{\mathcal{M}}$ are identical. Thus, $\mathcal{M}, s \models f$ iff $\bar{\mathcal{M}}, s \models f$.
2. Let $f := \sim f'$. From the induction hypothesis, $\mathcal{M}, s \not\models f'$ iff $\bar{\mathcal{M}}, s \not\models f'$. Thus, $\mathcal{M}, s \models f$ iff $\bar{\mathcal{M}}, s \models f$.
3. Let $f := f_1 \wedge f_2$. From the induction hypothesis, $\mathcal{M}, s \models f_1$ iff $\bar{\mathcal{M}}, s \models f_1$ and $\mathcal{M}, s \models f_2$ iff $\bar{\mathcal{M}}, s \models f_2$. Thus, $\mathcal{M}, s \models f$ iff $\bar{\mathcal{M}}, s \models f$.
4. Let $f := [\mathsf{X}f']_{\bowtie c}$. Consider the case $\bowtie \in \{\geq, >\}$. Suppose $\overset{*}{\eta}$ be the optimal scheduler of $\bar{\mathcal{M}}$ such that $Pr^{\overset{*}{\eta}}_{\bar{\mathcal{M}}}(\mathsf{X}f') = \min_{\eta} Pr^{\eta}_{\bar{\mathcal{M}}}(\mathsf{X}f')$.

   We show that for every ε we can construct a scheduler $\hat{\eta}$ of $\mathcal{M}$ such that

$$Pr^{\hat{\eta}}_{\mathcal{M}}(\mathsf{X}f') - Pr^{\overset{*}{\eta}}_{\bar{\mathcal{M}}}(\mathsf{X}f') \leq \varepsilon.$$

   Observe that, any scheduler of $\mathcal{M}$ is also a scheduler of $\bar{\mathcal{M}}$, since for any states $s, t \in S$ $\delta(s,t) \subseteq \bar{\delta}(s,t)$. Thus, Corollary 1. is applicable. Let $Q_s \triangleq \{t \mid \overset{*}{\eta}(s,t) > 0, \overset{*}{\eta}(s,t) \notin \delta(s)\}$ and $R_s \triangleq \{t \mid \overset{*}{\eta}(s,t) > 0, \overset{*}{\eta}(s,t) \in \delta(s,t)\}$. We assume that Q_s, R_s are not empty and there are no point intervals. Let $\hat{\eta}(s) = d$, where d is defined as follows:

- Let $t \in Q_s$. This implies $\overset{*}{\eta}(s,t) = \delta(s,t)\uparrow$ or $\overset{*}{\eta}(s,t) = \delta(s,t)\downarrow$. If $\overset{*}{\eta}(s,t) = \delta(s,t)\uparrow$ then $\delta(s,t)$ is open from above and $\boldsymbol{d}_t = \overset{*}{\eta}(s,t) - \frac{\varepsilon\rho}{|S|}$, where ρ is the minimum of the length of the non-zero interval in $\mathcal{M}$ and the $\overset{*}{\eta}(s,t)$ for $t \in R_s$. Similarly, if $\overset{*}{\eta}(s,t) = \delta(s,t)\downarrow$ then $\delta(s,t)$ is open from below and $\boldsymbol{d}_t = \overset{*}{\eta}(s,t) + \frac{\varepsilon\rho}{|S|}$.

- Let $t \in R_s$ and $\alpha \triangleq 1 - \sum_{t \in Q_s} \boldsymbol{d}_t - \sum_{t \in R_s} \overset{*}{\eta}(s,t)$. We have $\boldsymbol{d}_t = \overset{*}{\eta}(s,t) + \frac{\alpha}{|R_s|}$.

It follows that $\boldsymbol{d}$ is a distribution on the states of $\mathcal{M}$ and is a solution to $\delta(s)$. Let $E \triangleq \{w \mid \overset{*}{\eta}(w_0, w_1) > 0 \text{ and } \bar{\mathcal{M}}, w_1 \models f'\}$ and $E' \triangleq \{w \mid \hat{\eta}(w_0, w_1) > 0 \text{ and } \mathcal{M}, w_1 \models f'\}$.

$$| \underset{\bar{\mathcal{M}}}{\overset{\bar{\eta}}{\Pr}}(E) - \underset{\mathcal{M}}{\overset{\hat{\eta}}{\Pr}}(E')| \leq \sum_{t \in support(\hat{\eta}(s))} \frac{\varepsilon\rho}{|S|} \leq \varepsilon$$

Thus we can conclude that $\inf_\eta \Pr^\eta_{\mathcal{M}}(\mathsf{X}f') = \min_\eta \Pr^\eta_{\bar{\mathcal{M}}}(\mathsf{X}f')$. By similar argument:

$$\sup_\eta \Pr^\eta_{\mathcal{M}}(\mathsf{X}f') = \max_\eta \Pr^\eta_{\bar{\mathcal{M}}}(\mathsf{X}f').$$

$\mathcal{M}, s \models [\mathsf{X}f']_{\bowtie c}$ iff $\bar{\mathcal{M}}, s \models [\mathsf{X}f']_{\bowtie c}$, where $\bowtie \in \{\leq, <\}$.

5. Let $f := [f_1 \cup f_2]_{\bowtie c}$. Suppose $\bowtie \in \{\geq, >\}$. By induction hypothesis, for every s, $\mathcal{M}, s \models f_1$ iff $\bar{\mathcal{M}}, s \models f_1$ and $\mathcal{M}, s \models f_2$ iff $\bar{\mathcal{M}}, s \models f_2$. Let $S_1 \triangleq \{s \mid s, \mathcal{M} \models f_1\}$ and $T \triangleq \{s \mid s, \mathcal{M} \models f_2\}$. The IMC $\mathcal{M}'$ is obtained from $\mathcal{M}$ by omitting states not present in the set $S_1 \cup T$. It is easy to see that, if E is the event of reaching T in $\mathcal{M}'$, then $\inf_\eta \Pr^\eta_{\mathcal{M}'}(E) = \inf_\eta \Pr^\eta_{\mathcal{M}}(f)$. From Corollary 1 it follows that for any $0 < \varepsilon \leq 1$ we can find $\hat{\eta}$ such that $\Pr^{\hat{\eta}}_{\mathcal{M}'}(E) - \min_\eta \Pr^\eta_{\bar{\mathcal{M}}'}(E) \leq \varepsilon$, where E is the event of reaching T in $\mathcal{M}'$. Thus $\inf_\eta \Pr^\eta_{\mathcal{M}}(f) = \min_\eta \Pr^\eta_{\bar{\mathcal{M}}}(f)$. Similar argument holds for $\bowtie \in \{<, \leq\}$.

This concludes the proof.

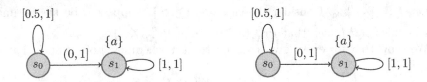

Fig. 3. A interval Markov chain   **Fig. 4.** A closed interval Markov chain

Example 4. Consider PCTL model checking of IMCs in UMC semantics. This involves existentially quantifying the transition probabilities and creating a formula in closed real field [4]. This captures a strict set of DTMC as compared to IMDP semantics, i.e, $[\mathcal{M}]_u \subsetneq [\mathcal{M}]_d$. For example, DTMC where the transition probability between two states s, t change over time cannot be represented in UMC semantics. This is exemplified by the IMC $\mathcal{M}$ in Figure 3. The probability of satisfying the path formula $g = \mathsf{G} \ (\sim a \wedge [\mathsf{X}a]_{>0})$ in the UMC semantics is 0. But we can find schedulers which can make the probability of satisfying g arbitrarily close to 1. The scheduler has the freedom to define an infinite Markov chain by assigning monotonically increasing probabilities for the transition $s_0 \rightarrow s_0$).

The model checking of the open IMC $\mathcal{M}$ is done by closing it (Figure 4). This gives us the closed IMC $\bar{\mathcal{M}}$, shown below: The maximum probability of satisfying g in $\bar{\mathcal{M}}$ is 1. Which implies, for every $0 < \varepsilon \leq 1$, there exists a scheduler $\hat{\eta}$, for which the probability of staying in a state that satisfies $\sim a \wedge [\mathsf{X}a]_{>0}$ (s_0) is greater than $1 - \varepsilon$, by Theorem 2.

6   Conclusion

We presented the problem of model checking Interval Markov chains with open intervals. We proved that as far as model checking (and reachability) is concerned open intervals do not cause any problem in interval Markov decision process semantics and thus can be safely ignored. Interval Markov chains are but special cases of more complex Markovian models, called *constraint Markov chains* (CMC) [3]. Transition probabilities in these models are defined as a solution to linear equations. Let F_V be the set of linear in-equations on variables V. A constraint Markov chain is a tuple $\mathcal{M} \triangleq (S, L, \delta)$, where the transition function $\delta : S \rightarrow 2^{F_V}$, maps each state to a set of linear in-equations. Thus IMCs are a strict sub-class of convex Markov decision process. The behaviour of a CMC can be defined in the UMC and IMDP semantics. We say, a system of in-equation are closed if they have non-strict inequalities, otherwise they are open. A CMC is called open if the transition function maps to an open system of linear equations. Model checking open CMCs have the same kinds of problems as described for IMCs. Theorem 2. can be extended to CMCs as well. We can define *basic feasible solutions* for a system of linear in-equations as well. Let s be a state of a CMC $\mathcal{M}$ and $\delta(s)$ be a system of linear in-equations on variables $\{x_1, \ldots, x_k\}$ such that x_i denotes the probability of moving from state s to s_i. The BFSs of $\delta(s)$ are the vertices of the convex hull defined by the set of in-equations $\delta(s) \cup \{x_1 + \ldots + x_k = 1\}$. The same argument as in the proof of Theorem 2 shows that, model checking of PCTL formulas on CMCs can be done by first closing the system of in-equations, this is done by replacing the strict inequalities $(<, >)$ with non-strict inequalities $(\leq, \geq)$, and then model checking on the closed model.

Acknowledgement. The authors thank Hongfei Fu for discussions on the topic of this paper.

References

1. Baier, C., Katoen, J.-P.: Principles of Model Checking (Representation and Mind Series). The MIT Press (2008)
2. Bianco, A., De Alfaro, L.: Model checking of probabilistic and nondeterministic systems. In: Thiagarajan, P.S. (ed.) FSTTCS 1995. LNCS, vol. 1026, pp. 499–513. Springer, Heidelberg (1995)
3. Caillaud, B., Delahaye, B., Larsen, K.G., Legay, A., Pedersen, M.L., Wasowski, A.: Constraint Markov chains. Theoretical Computer Science **412**(34), 4373–4404 (2011)
4. Chatterjee, K., Henzinger, T., Sen, K.: Model-checking omega-regular properties of interval Markov chains. In: Amadio, R.M. (ed) Foundations of Software Science and Computation Structure (FoSSaCS) 2008, pp. 302–317, March 2008
5. Clarke Jr., E.M., Grumberg, O., Peled, D.A.: Model Checking. MIT Press, Cambridge (1999)
6. Courcoubetis, C., Yannakakis, M.: The complexity of probabilistic verification. J. ACM **42**(4), 857–907 (1995)
7. Allen Emerson, E.: Temporal and modal logic. In: Handbook of Theoretical Computer Science (vol. b), pp. 995–1072. MIT Press, Cambridge (1990)
8. Hansson, H., Jonsson, B.: A logic for reasoning about time and reliability. Formal Aspects of Computing **6**(5), 512–535 (1994)
9. Jonsson, B., Larsen, K.G.: Specification and refinement of probabilistic processes. In: LICS, pp. 266–277. IEEE Computer Society (1991)
10. Renegar, J.: On the computational complexity and geometry of the first-order theory of the reals. part i: Introduction. preliminaries. the geometry of semi-algebraic sets. the decision problem for the existential theory of the reals. Journal of Symbolic Computation **13**(3), 255–299 (1992)
11. Sen, K., Viswanathan, M., Agha, G.: Model-checking Markov Chains in the presence of uncertainties. In: Hermanns, H., Palsberg, J. (eds.) TACAS 2006. LNCS, vol. 3920, pp. 394–410. Springer, Heidelberg (2006)
12. Walley, P.: Measure of uncertainty in expert systems. Artificial Intelligence **83**(1), 1–58 (1996)
13. Škulj, D.: Discrete time Markov Chains with interval probabilities. International Journal of Approximate Reasoning **50**(8), 1314–1329 (2009). Special Section on Interval/Probabilistic Uncertainty

Performance Modeling of Cellular Systems with Finite Processor Sharing Queues in Random Environment, Guard Policy and Flex Retrial Users

Ioannis Dimitriou[✉]

Department of Mathematics, University of Patras, 26500 Patras, Greece
idimit@math.upatras.gr

Abstract. We investigate a two-station retrial queueing system to model the access in modern cellular networks managed by two service providers. Each provider owns a single access point, which operates under processor sharing discipline, and accepts three types of users: the handover users and, the originating subscribers and the originating flex users. At the arrival epoch, a flex user connects with the provider, which offers the largest data rate. Each access point can admit a limited number of users and employ a guard bandwidth policy in order to prioritize the handover users. Both blocked handover and originating subscriber users are lost. Blocked flex users join a virtual orbit queue of infinite capacity from where they retry independently to connect with the service provider that offers the largest data rate at their retrial time. Moreover, the system operates in varying environmental conditions. Using the matrix analytic formalism we construct a four-dimensional Markovian model, which allows to represent accurately the types of user behavior and the environmental aspects in cellular networks. We perform a steady-state analysis and a study of the main performance metrics.

Keywords: Finite processor sharing · Retrials · Load balancing · Matrix analytic method · Random environment

1 Introduction

In modern cellular mobile networks, the importance of a proper modeling of customer behavior for improving the quality of service provided by the system has been stressed [1,11,37]. Such systems are characterized by the partitioning of the coverage area into cells served by a base station that can handle a limited number of users. Typically the system must handle not only the fresh sessions (i.e. calls) initiated inside a given cell, but also the handover sessions of users moving across the cell.

An important issue is the modeling of the handover session process, which consists of session requests caused by mobile users moving from one cell to another. The current ongoing session has to be handed over between base stations. Taking into account the customer mobility and the handover effect, the

© Springer International Publishing Switzerland 2015
M. Gribaudo et al. (Eds.): ASMTA 2015, LNCS 9081, pp. 43–58, 2015.
DOI: 10.1007/978-3-319-18579-8_4

cell faces two kinds of session arrival processes: fresh sessions, i.e. sessions originated in that cell, and handovers sessions. Since handover users already use network resources, they should be completed first. Normally they are prioritized with respect to fresh sessions, since blocking a handover user, will degrade more seriously the quality of service (QoS). Moreover, for an accurate representation of a cellular network, it cannot be ignored that blocked sessions are able to redial after some random time. Several studies [1, 9–11, 24, 26, 37] recognize the important role played by the retrial phenomenon. For an account of the main results on retrial queues, we refer the reader to the monographs [2, 12].

The time spent by users before starting a retrial with modern mobile handsets becomes shorter compared to the conventional telephone systems. These retrials will have a negative influence on fresh sessions being connected at their first attempt, and on handover requests, as the offered load of the system becomes higher. To cope with the problem of blocked handovers, which must be avoided, the concept of guard policy has been proposed.

Under such a a scheme, network resources can be reserved in individual cells or form a pool for a reuse area, and can only be used by handover users. One of the first studies on this subject is given by Guerin [15] (see also [6]), who explicitly studied a model with no retrial users. We also mention the nominal papers by [7], [37], and [33] that introduced the guard channel policy in the retrial context (see also [9, 10] where guard and fractional guard policy was investigated in conjunction with user retrials). Recently, Kim et al. [20] provided a detailed analysis of stationary queue length distribution along with an optimization formulation for the guard policy in the presence of retrials. In another scheme, a priority queue is provided only to handover users [8, 17].

It is well known that real communication systems do not have stationary arrival and/or service patterns. In contrast, queueing parameters vary randomly over time due to a variety of reasons including rush hour behavior, existence of several customer classes, breakdowns and repairs of the servers, and random shocks (see, e.g. [18, 28, 35]). However, the joint consideration of the retrial feature and the random environment has been addressed only in a few papers [3, 18, 19, 35]. Clearly, all the studies mentioned above refer to a single station.

Furthermore, the Processor Sharing discipline (PS) has been used to evaluate the flow-level performance of cellular data systems using Proportional Fairness scheduling [4]. It is a popular model in the study of bandwidth sharing on the Internet [14]. Some researchers also use it in the performance analysis of wireless LANs [25]. In an egalitarian PS queue, the servers capacity[1] is shared equally among all flows concurrently in service. In addition, in such systems, the transmission of a blocked flow is re-tried after some rethinking time with a certain probability [21].

In this paper we focus on the effect of the retrial phenomenon on the QoS in a cellular mobile network as well as the introduction of flex users that have a special treatment and choose to connect with the service station that offers the largest

---

[1] Server capacity may mean transmission slots, bandwidth, or CPU time, depending on the system being studied.

data rate upon arrival. Recent studies on game theoretical analysis of cellular markets [13], reveal the significant benefits of the notion of flex service compared to traditional subscriber-only markets. It was shown that the flex service reduces the percentage of disconnected users, and improves the social welfare. More importantly, flex users exhibited significantly lower blocking probabilities than subscribers.

In the following, we consider a mobile communication system operating in a random Markovian environment and focus on a single cell served by two access points (APs) that belong to two different service providers (SPs). Each service station (i.e., an AP of a SP) offers a specific bandwidth, which is shared by all connected users (i.e., a PS service discipline), and can handle a limited number of users (see [22,30]). Moreover, we let, for the first time in the related literature, the service rate to depend both on the state of the environmental process, and on the number of the connected users. Each AP handles three types of customers, handover users, originating subscribers and originating flex users.

A guard bandwidth policy [36] is applied to each service station in order to give priority to handover users. More precisely, given the access point's bandwidth, say C, its admission region in the cell is given by (N_c, G_h, G_f), where $N_c = \lfloor C/r \rfloor$ is the maximum number of connected users, where r is the the minimum bandwidth requirement of a call, G_f ($\leq N_c$) is a real number representing a randomized number of guard channels (guard bandwidth) dedicated to new and handoff traffic in the cellular area, and G_h ($\leq G_f$) is the guard bandwidth reserved only for handoff traffic in this area. Both blocked handover and originating subscribers users are considered lost. For the first time in the retrial literature, originating flex users join the AP that provides the largest data rate at the arrival moment.

If both APs are fully utilized, originating flex users join an infinite capacity orbit and retry for service. A retrial flex user will try to connect with the AP that offers the largest data rate upon retrial instant. Note that in such a case, the retrial rate of the flex user depends on the number of already connected users in each AP. Note also that it is the first time in the retrial queueing literature that a processor sharing discipline is used along with a guard policy, the presence of multi-station queueing system, the random environment, and more importantly the presence of flex users that join the AP which provides the largest data rate upon arrival.

To conclude, we can consider the following interpretations to understand the applicability of flex service. In the first one, flex users may have a special type of contract that guarantees the coalition of both SPs. This special contract allows a flex user to connect with the SP which offers the largest data rate. Secondly, flex users may have mobile devices that are equipped with a special user centric application which provides information in real time, about the offered data rate from the available SPs in the target cell, and forces the user to connect with the "best" one.

We show how the matrix analytic formalism [23,28] provides an appropriate mathematical framework to construct a four-dimensional Markovian model which allows us to represent accurately the types of user behavior and the environmental aspects in cellular mobile networks. The organization of the rest of this paper is as follows. In Section 2, we describe the mathematical model. The construction of the underlying block-structured infinitesimal generator is presented in Section 3, while in Section 4, the stability condition and the stationary distribution of the system state is derived. Section 5 deals with the derivation of formulas for some key performance measures of the system. Finally, in Section 6, we give a numerical example that illustrates the system performance.

2   Model Description

We consider a queueing model composed of two stations that operate under processor sharing discipline. Each station has a finite capacity and cannot admit more than a specific number of connected users. Therefore, we consider a time-sharing system of two queues that admit at most M_1, M_2 users respectively. The queueing system accepts five types of users, say P_{1h}, P_{2h}, P_{1f}, P_{2f} and P_{3f}. P_{kf} users, $k = 1, 2$ are subscribers that generate sessions originated in the target cell (fresh sessions) and they connect only with the service provider k. P_{3f} users are flex (originated sessions in the target cell) in the sense that connect to a provider, which offers the largest data rate (see below for details). Moreover, P_{1h}, P_{2h} are handover users arriving from adjoining cells, already connected with the SP 1, 2 respectively.

The limited bandwidth of a target cell and the competition between the users may create essential problems, especially for the moving users. When an active mobile user enters the target cell moving from the adjoining cell, his/her communication can be terminated due to lack of free resources. The requests of such on-going (handover) users compete with the requests of the users originated in the target cell (fresh sessions). Clearly, it is more intolerable to drop an on-going service, than to block a service that has yet to be established. In order to provide some kind of priority to handover users, we employ a guard policy [36], which assumes a reservation of AP's bandwidth exclusively for the service of handover users.

Let R_j be a real number representing a randomized number of guard channels (guard bandwidth or simply service positions) dedicated only for handover users in AP j, $j = 1, 2$. That said, let $C_j^{(i)}$ is the offered bandwidth by the AP j, given the environmental state i (see below). We reserve a part $C_{jh}^{(i)} < C_j^{(i)}$ for handover requests. Thus, $R_j = C_{jh}^{(i)}/r_-^{(i)}$, where $r_-^{(i)}$ is the minimum attainable transfer rate, given the environmental state i (Note that for convenience we have assumed that $M_j = C_j^{(i)}/r_-^{(i)}$, $j = 1, 2$, $\forall i$). Therefore, $L_j = M_j - R_j = (C_j^{(i)} - C_{jh}^{(i)})/r_-^{(i)}$ service positions can be shared from the fresh users originated in the target cell and from handover users.

The behavior of the model depends on the state of the environmental conditions, which are governed by an irreducible continuous time Markov chain $\{Y(t); t \geq 0\}$ with with finite state space $E = \{1, 2, ..., M\}$ and infinitesimal generator S. The environmental process is a very useful modeling tool that offers us the ability to adapt in our model the fluctuation of users' traffic. Handover users connected with SP k, arrive according to a Poisson process with rate $\lambda_{kh}^{(i)}$. If they find available space in the AP k, will occupy a service position, otherwise the handover user is lost and so the mobile user during a conversation is forced to be terminated. Moreover, P_{kf} users arrive at the system according to a Poisson process with rate $\lambda_k^{(i)}$, $k = 1, 2, 3$, given that the environmental process is in state i. Upon arrival, P_{kf}, $k = 1, 2$, users join AP k, if there is available space (i.e., at least L_k available service positions) at the corresponding station.

We assume that each user has an exponentially distributed service requirement with mean $1/\mu$. When the environmental process is in state i, and there are m_k connected users in AP k, the server works at a rate $c_{m_k}^{(i)} > 0$. This service capacity is equally shared among all connected users. Hence, each user terminates his session in an interval of length Δ with probability $\frac{1}{m_k}\mu c_{m_k}^{(i)}\Delta + o(\Delta)$, for $\Delta \to 0$. Let $\mu_{km_k}^{(i)} = \mu c_{m_k}^{(i)}$. Note that under such scheme we can incorporate unavailability periods for the APs. That said, we can set some of the $c_{m_k}^{(i)}$ to be zero. In practical applications this fact depends on the state of the environmental state i. For instance, it is possible during the weekend the AP to be unavailable due to maintenance (see [30, 31]). Moreover, the speed $c_{m_k}^{(i)}$ depends on the total number of connected users and thus, we may write $c_{m_k}^{(i)} = min[r_+^{(i)}, C_k^{(i)}/m_k]$, where $r_+^{(i)} \geq r_-^{(i)}$ the maximum attainable transfer rate.

Upon arrival, a P_{3f} user joins the station which provides the largest data rate. That said, he/she chooses AP k, where $\mu_{km_k+1}^{(i)}/(m_k+1) = max(\mu_{1m_1+1}^{(i)}/(m_1+1), \mu_{2m_2+1}^{(i)}/(m_2+1))$. If both APs have the same occupancy, the arriving P_{3f} customer is routed randomly to each AP with probability $1/2$, provided that there is available space in both APs. If an AP is fully utilized, the arriving user will connect to the AP which has a vacant place.

Moreover, we assume that the dwelling times (i.e., the time spent in a cell before moving to the coverage area of an adjacent cell) of users are also exponentially distributed with rate ν_i, independent of the type of user given that the environmental process is in state i, $i = 1, ..., M$. Therefore, the total holding time for a connected user in AP k is exponential with rate $\mu_{km_k}^{(i)} + m_k\nu_i$, $m_k = 1, ..., M_k$. On the other hand, blocked flex users abandon the cell due to their mobility at a rate $n\nu_i$, given that there are n such users.

Letting the handover rate ν_i (i.e., the rate of dwell time) to depend on the environmental state, we can deal with several practical issues. For example, consider the case of smart APs, operating by solar panels (see Mancuzo and Alouf [27]) that dynamically adjust their coverage area according to weather conditions. Assume that the weather alternates between sunny and cloudy (i.e., $M = 2$, $i = 1, 2$). When the weather is sunny ($i = 1$), then the APs have enough energy capacity to operate efficiently and they increase their coverage area. On the other hand, if it is cloudy ($i = 2$) they decrease their coverage area in order to save energy. In such a case, $\nu_1 < \nu_2$, i.e., the user spends more time in the cell during sunny periods, compared with the case of cloudy periods. An alternative example will allow fluctuation in the velocity of the mobile users, and thus, the time spent by a user in a cell will vary according to the environmental state.

Since each AP can serve a limited number of users, blocking phenomena occur. Find below the blocking rules:

1. A handover user, connected in SP j, is rejected (and lost) if $M_j = L_j + R_j$ service positions are occupied in AP j, at an arrival instant.
2. A fresh session originating by a subscriber of SP j is blocked and lost if at least L_j service positions are occupied.
3. If an arriving fresh P_{3f} user finds both APs fully occupied, enters a virtual queue (i.e., orbit) of infinite capacity from which try, independently of each other to access the system after an exponentially distributed time period with rate $\alpha^{(i)}$, $i = 1, ..., M$. Upon a retrial attempt, a flex user will join an AP under the maximum data rate policy described above, given that both APs have less than L_j, $j = 1, 2$, occupied service positions. If an AP has no service positions available for originating users then the retrial flex user will connect to the other AP. Otherwise, the user will retry later on.

For later use, define $\Lambda_k = diag(\lambda_k^{(1)}, ..., \lambda_k^{(M)})$, $\Lambda_{kh} = diag(\lambda_{kh}^{(1)}, ..., \lambda_{kh}^{(M)})$, $k = 1, 2$ and $\Lambda_3 = diag(\lambda_3^{(1)}, ..., \lambda_3^{(M)})$, $\underline{\alpha} = (\alpha^{(1)}, ..., \alpha^{(M)})$.

Remark 1. Our model is general enough to describe many practical situations. For example, consider two Web server farms, where the scheduling of jobs at the hosts is modeled by PS discipline [16]. In such a case, flex users may represent jobs that have a specific Service Level Agreement with both SPs, that guarantees the coalition between them and as a result, these jobs will be served by the service station, which provides the largest service rate at the arrival instant.

Remark 2. The matrix analytic approach that we use in the following sections, is powerful enough to describe the problem of $k > 2$ APs that serve the target cell. However, the computational cost to analyze the underlying stochastic process will increase rapidly, and thus, in such a case, maybe a simulation study is preferred.

3   Process of the System States

Let $Q_r(t)$, $Q_j(t)$, and $Y(t)$ denote, respectively, the number of blocked flex users in orbit, the number of connected users in AP j, $j = 1, 2$ and the state of the environmental process at time t. The orbit is of infinite capacity. The process $U = \{(Q_r(t), Q_1(t), Q_2(t), Y(t)); t \geq 0\}$ describes our system (see Fig. 1) and is a continuous-time Markov chain, and in particular a Level Dependent Quasi Birth Death (LDQBD) process. The state space of U is given by $H = \cup_{n=0}^{\infty} l(n)$, where the subsets $l(n)$, $n \geq 0$ are the levels of the LDQBD given by $l(n) = \{(n, m_1, m_2, i), 0 \leq m_j \leq M_j, j = 1, 2, 0 \leq i \leq M\}$.

Table 1. Overview of system's parameters

Description	Value
Offered (reserved) bandwidth at AP j:	$C_j^{(i)}$ ($C_{jh}^{(i)}$), $i = 1, ..., M$, $j = 1, 2$.
Minimum/maximum attainable transfer rate of a call:	$r_{-/+}^{(i)}$, $i = 1, ..., M$
Maximum number of connected users in AP j:	$M_j = C_j^{(i)}/r_-^{(i)}$, $j = 1, 2$.
Number of states of the environmental process:	M.
Arrival rate of handover users at AP j:	$\lambda_{jh}^{(i)}$, $i = 1, ..., M$, $j = 1, 2$
Arrival rate of originating subscribers at AP j:	$\lambda_j^{(i)}$, $i = 1, ..., M$, $j = 1, 2$
Arrival rate of originating flex users:	$\lambda_{3f}^{(i)}$, $i = 1, ..., M$
The rate of the dwelling time:	ν_i, $i = 1, ..., M$
Retrial rate of blocked flex users:	$\alpha^{(i)}$, $i = 1, ..., M$
Randomized number of guard channels of AP j:	$R_j = C_{jh}^{(i)}/r_-^{(i)}$, $j = 1, 2$
Total service rate at AP j:	$\mu_{jm_j}^{(i)} = \mu c_{m_j}^{(i)}$, $i = 1, ..., M$, $j = 1, 2$

Then, the infinitesimal generator Q, in partitioned form, is given by

$$Q = \begin{pmatrix} Q_{0,0} & C & 0 & 0 & \cdots \\ Q_{1,0} & Q_{1,1} & C & 0 & \cdots \\ 0 & Q_{2,1} & Q_{2,2} & C & \cdots \\ \vdots & \vdots & \vdots & \ddots & \ddots & \ddots & \vdots \\ \vdots & \vdots & \vdots & \vdots & \ddots & \ddots & \ddots \end{pmatrix},$$

where, $Q_{i,j}$ are square matrices of order $M(M_1+1)(M_2+1)$. Moreover, for $n \geq 0$,

$$Q_{n,n} =$$

$$\begin{pmatrix} Q_{n,n}^{(0,0)} & Q_{n,n}^{(0,1)} & 0 & 0 & \cdots \\ Q_{n,n}^{(1,0)} & Q_{n,n}^{(1,1)} & Q_{n,n}^{(1,2)} & 0 & \cdots \\ 0 & Q_{n,n}^{(2,1)} & Q_{n,n}^{(2,2)} & Q_{n,n}^{(2,3)} & \cdots \\ \vdots & \vdots & \ddots & \ddots & \ddots & \vdots \\ \vdots & \vdots & \vdots & \vdots & \ddots & \ddots \\ & & & & Q_{n,n}^{(M_1-1,M_1-2)} & Q_{n,n}^{(M_1-1,M_1-1)} & Q_{n,n}^{(M_1-1,M_1)} \\ & & & & & Q_{n,n}^{(M_1,M_1-1)} & Q_{n,n}^{(M_1,M_1)} \end{pmatrix},$$

where, for $m_1 = 1, ..., M_1$,

$$\begin{aligned} Q_{n,n}^{(m_1,m_1-1)} &= I_{(M_2+1)\times(M_2+1)} \otimes M_1^{(m_1)}, \\ Q_{n,n}^{(m_1,m_1+1)} &= I_{(M_2+1)\times(M_2+1)} \otimes \Lambda_1^{(m_1,m_2)}, \end{aligned} \tag{1}$$

and for $k = 1, 2$,

$$\begin{aligned} M_k^{(m_k)} &= diag(\mu_{km_k}^{(1)} + m_k\nu_1, ..., \mu_{km_k}^{(M)} + m_k\nu_M), \\ \widehat{\Lambda}_k^{(m_1,m_2)} &= diag(\Lambda_{k1}^{(m_1,m_2)}, ..., \Lambda_{kM}^{(m_1,m_2)}). \end{aligned} \tag{2}$$

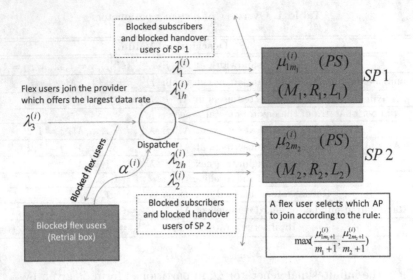

Fig. 1. The model given that the environmental process is in state i

Denote $u_{km_k}^{(i)} = \frac{\mu_{km_k}^{(i)}}{m_k}$, $k = 1, 2$, $i = 1, ..., M$. Then,

$$
\begin{aligned}
\Lambda_{1i}^{(m_1,m_2)} &= \lambda_{1h}^{(i)} \delta_{\{m_1 < M_1\}} + [\lambda_1^{(i)} + \lambda_3^{(i)} \delta_{\left\{u_{1m_1+1}^{(i)} > u_{2m_2+1}^{(i)}\right\}} \\
&\quad + \frac{\lambda_3^{(i)}}{2} \delta_{\left\{u_{1m_1+1}^{(i)} = u_{2m_2+1}^{(i)}\right\}}] \delta_{\{m_1 < L_1\}}, \\
\Lambda_{2i}^{(m_1,m_2)} &= \lambda_{2h}^{(i)} \delta_{\{m_2 < M_2\}} + [\lambda_2^{(i)} + \lambda_3^{(i)} \delta_{\left\{u_{1m_1+1}^{(i)} < u_{2m_2+1}^{(i)}\right\}} \\
&\quad + \frac{\lambda_3^{(i)}}{2} \delta_{\left\{u_{1m_1+1}^{(i)} = u_{2m_2+1}^{(i)}\right\}}] \delta_{\{m_2 < L_2\}}.
\end{aligned}
\tag{3}
$$

Furthermore, for $m_1 = 0, 1, ..., M_1 - 1$,

$$
Q_{n,n}^{(m_1,m_1)} =
$$

$$
\begin{pmatrix}
T_{00}^{(n)} & \widehat{\Lambda}_2^{(m_1,0)} & 0 & 0 & 0 & \cdots & & & \\
M_2^{(1)} & T_{11}^{(n)} & \widehat{\Lambda}_2^{(m_1,1)} & 0 & 0 & \cdots & & & \\
0 & M_2^{(2)} & T_{22}^{(n)} & \widehat{\Lambda}_2^{(m_1,2)} & 0 & \cdots & & & \\
\vdots & \vdots & \vdots & \ddots & \ddots & \ddots & \vdots & & \\
\vdots & \vdots & \vdots & \vdots & \ddots & \ddots & \ddots & & \\
 & & & & & M_2^{(M_2-1)} & T_{M_2-1\,M_2-1}^{(n)} & \widehat{\Lambda}_2^{(m_1,M_2-1)} \\
 & & & & & & M_2^{(M_2)} & T_{M_2 M_2}^{(n)}
\end{pmatrix},
$$

where,

$$T_{m_2 m_2}^{(n)} = S - \Lambda_3 - H^{(n)} - \sum_{k=1}^{2} [\Lambda_k \delta_{\{m_k < L_k\}} + \Lambda_{kh} \delta_{\{m_k < M_k\}} + A_k^{(m_1, m_2, n)} + M_k^{(m_k)}],$$

and $H^{(n)} = diag(n\nu_1, ..., n\nu_M)$, $A_k^{(m_1, m_2, n)} = diag(\alpha_{k1}^{(m_1, m_2, n)}, ..., \alpha_{kM}^{(m_1, m_2, n)})$,

$$\alpha_{1i}^{(m_1, m_2, n)} = n[\alpha^{(i)} \delta_{\{u_{1m_1+1}^{(i)} > u_{2m_2+1}^{(i)}\}} + \frac{\alpha^{(i)}}{2} \delta_{\{u_{1m_1+1}^{(i)} = u_{2m_2+1}^{(i)}\}}] \delta_{\{m_1 < L_1\}},$$
$$\alpha_{2i}^{(m_1, m_2, n)} = n[\alpha^{(i)} \delta_{\{u_{1m_1+1}^{(i)} < u_{2m_2+1}^{(i)}\}} + \frac{\alpha^{(i)}}{2} \delta_{\{u_{1m_1+1}^{(i)} = u_{2m_2+1}^{(i)}\}}] \delta_{\{m_2 < L_2\}}. \tag{4}$$

Matrix C is of size $M \prod_{j=1}^{2} (M_j + 1)$ and describes transitions between the levels of the QBD that correspond to the arrival of a flex user in the orbit queue. Note that, a flex user enters the orbit queue when it finds upon arrival at least L_1 and L_2 connected users in the AP 1, 2 respectively. Then,

$$C = \begin{pmatrix} 0_{ML_1(M_2+1) \times ML_1(M_2+1)} & 0_{ML_1(M_2+1) \times M(M_1-L_1+1)(M_2+1)} \\ 0_{M(M_1-L_1+1)(M_2+1) \times ML_1(M_2+1)} & I_{(M_1-L_1+1) \times (M_1-L_1+1)} \otimes F \end{pmatrix},$$

where

$$F = \begin{pmatrix} 0_{ML_2 \times ML_2} & 0_{ML_2 \times M(M_2-L_2+1)} \\ 0_{M(M_2-L_2+1) \times ML_2} & I_{M(M_2-L_2+1) \times M(M_2-L_2+1)} \otimes \Lambda_3 \end{pmatrix}.$$

Finally, for $n \geq 1$, $Q_{n,n-1} = (Q_{(n,m_1),(n-1,m_1')})$, where $Q_{(n,m_1),(n-1,m_1')} = 0_{M(M_2+1) \times M(M_2+1)}$,

$$Q_{(n,m_1),(n-1,m_1))} = L_{m_1 m_1}^{(n)}, m_1 = 1, ..., M_1,$$
$$Q_{(n,m_1),(n-1,m_1+1))} = L_{m_1 m_1+1}^{(n)} \tag{5}$$
$$= diag(A_1^{(m_1,0,n)}, A_1^{(m_1,1,n)}, ..., A_1^{(m_1,M_2,n)}), m_1 = 1, ..., M_1 - 1.$$

The sub-blocks $L_{m_1 m_1}^{(n)} = (L_{(m_1,m_2),(m_1,m_2')}^{(n)})$ are of order $M(M_2+1) \times M(M_2+1)$, where $L_{(m_1,m_2),(m_1,m_2')}^{(n)} = 0_{M \times M}$, $m_2' - m_2 > 1$, $m_2' - m_2 \leq -1$, and

$$L_{(m_1,m_2),(m_1,m_2)}^{(n)} = H^{(n)}, L_{(m_1,m_2),(m_1,m_2+1)}^{(n)} = A_2^{(m_1,m_2,n)}. \tag{6}$$

4 Stationary Distribution

Let $\underline{x}$, partitioned as $\underline{x} = (\underline{x}_0, \underline{x}_1, ...)$, $\underline{x}_n = (\underline{x}_{n0}, \underline{x}_{n1}, ..., \underline{x}_{nM_1})$, $n \geq 0$,

$$\underline{x}_{nm_1} = (\underline{x}_{nm_10}, ..., \underline{x}_{nm_1M_2}), 0 \leq m_1 \leq M_1,$$
$$\underline{x}_{nm_1m_2} = (x_{n,m_1,m_2,1}, ..., x_{n,m_1,m_2,M}) 0 \leq m_j \leq M_j, j = 1, 2, \tag{7}$$

be the stationary probability vector satisfying

$$\underline{x}Q = 0, \underline{x}\underline{1}' = 1, \tag{8}$$

where 0 and $\underline{1}'$ denote a row vector and a column vector of zeros and ones with an appropriate size, respectively.

Several truncation methods have been proposed for solving the set of equations (8): (i) Direct truncation method [12], where the maximum orbit size is fixed to N^*. This method results in a finite level-dependent QBD with $N^* + 1$ levels, (ii) Generalized truncation method [29], where the retrial rate from the orbit is fixed by $N^* \alpha^{(i)}$ when there are $k \geq N^*$ customers present in the orbit. Neuts [29] indicated that the generalized truncation method is better that the direct truncation method, since the former requires a smaller truncation point than the latter does, in order to achieve the same accuracy.

Clearly, is not our aim here to provide a comparison between different methods for computing the stationary distribution. We use Neuts' approximation method [29] and assume that only N^* users among the blocked flex retrial users in the orbit can retry for the service even if there are retrial users greater than N^* (A choice of appropriate value of N^* is done following [29]). Then, the infinitesimal generator Q^* is modified as,

$$
Q^* = \begin{pmatrix}
Q_{0,0} & C & & & & & & \\
Q_{1,0} & Q_{1,1} & C & & & & & \\
& \ddots & \ddots & & \ddots & & & \\
& & Q_{N^*-1,N^*-2} & Q_{N^*-1,N^*-1} & C & & & \\
& & & Q_{N^*,N^*-1} & Q_{N^*,N^*} & C & & \\
& & & & Q_{N^*-1,N^*} & Q_{N^*,N^*} & C & \\
& & & & & \ddots & \ddots & \ddots
\end{pmatrix}.
$$

Let the stationary probability vector $\underline{\pi}$ (a $1 \times M(M_1+1)(M_2+1)$ vector) of the generator $A_{N^*} = C + Q_{N^*,N^*} + Q_{N^*,N^*-1}$. The vector $\underline{\pi} = (\underline{\pi}(0), ..., \underline{\pi}(M_1))$, where $\underline{\pi}(m_1) = (\pi(m_1, 0, 1), ..., \pi(m_1, M_2, M))$. Then,

Theorem 1. *The modified QBD with infinitesimal generator Q^* is positive recurrent if and only if*

$$
\sum_{m_1=L_1}^{M_1} \underline{\pi}(m_1)(I_{(M_1-L_1+1)\times(M_1-L_1+1)} \otimes F)\underline{1}' < \underline{\pi} Q_{N^*,N^*} \underline{1}'. \tag{9}
$$

Proof: The proof is done following the lines in [28].   □

Under the condition (9), the partitioned stationary probability vector $\underline{x}^*$ of Q^* is given by

$$
\underline{x}_n^* = \underline{x}_{N^*-1}^* R^{n-N^*+1}, \; n \geq N^* - 1, \tag{10}
$$

where the rate matrix R is the minimal non negative matrix solution of

$$
R^2 Q_{N^*,N^*-1} + R Q_{N^*,N^*} + C = 0_{M(M_1+1)(M_2+1)\times M(M_1+1)(M_2+1)}, \tag{11}
$$

with maximal eigenvalue $sp(R) < 1$, and the vectors $\underline{x}_n^*$, $0 \le n \le N^* - 1$, are obtained by the following set of equations

$$
\begin{aligned}
&\underline{x}_0^* Q_{0,0} + \underline{x}_1^* Q_{1,0} = \underline{0}, \\
&\underline{x}_{n-1}^* C + \underline{x}_{n-1}^* Q_{n,n} + \underline{x}_{n+1}^* Q_{n+1,n} = \underline{0}, \ 1 \le n \le N^* - 2, \\
&\underline{x}_{N^*-2}^* C + \underline{x}_{N^*-1}^* (Q_{N^*-1,N^*-1} + R Q_{N^*,N^*-1}) = \underline{0},
\end{aligned}
\tag{12}
$$

subject to the normalizing condition,

$$
\sum_{n=0}^{N^*-2} \underline{x}_n^* \underline{1}' + \underline{x}_{N^*-1}^* (I - R)^{-1} \underline{1}' = 1.
$$

Regarding the computation of the rate matrix, Bright and Taylor [5] (see also [23]) proposed an efficient algorithm for level-dependent QBDs with infinitely many levels. Recently, Phung-Duc et al. [32] developed a simple direct-truncated method in order to compute it, which is less memory consuming than Bright and Taylor [5]'s algorithm. In our case the matrix R is approximated by the following iteration

$$
R(0) = 0, \ R(l+1) = (C + R^2(l) Q_{N^*,N^*-1})(-Q_{N^*,N^*})^{-1}, \ l \ge 0,
\tag{13}
$$

and iterations will be continued until $max_{ij}[R_{ij}(l+1) - R_{ij}(l)] < \epsilon$, where $R(l)$ is the lth iteration and $\epsilon = 10^{-14}$ is the degree of the required accuracy.

5   Performance Metrics

Once the stationary probabilities have been computed, we can easily find the main stationary system performance characteristics.

1. The mean rate of arriving users

$$
\begin{aligned}
\Lambda = {} & \sum_{i=1}^{M} \sum_{m_1=0}^{M_1} \sum_{m_2=0}^{M_2} \sum_{n=0}^{N^*-1} s_i (\lambda_h^{(i)} + \lambda^{(i)} + n\alpha^{(i)}) x_{n,m_1,m_2,i}^* \\
& + \underline{x}_{N^*-1}^* R (I - R)^{-1} \underline{s}[N^* \underline{\alpha} + \underline{\lambda}_h + \underline{\lambda}]',
\end{aligned}
\tag{14}
$$

where, $\underline{s} = (s_1, ..., s_M)$ the invariant vector of the random environment, $\lambda_h^{(i)} = \sum_{k=1}^{2} \lambda_{kh}^{(i)}$, $\lambda^{(i)} = \sum_{k=1}^{3} \lambda_k^{(i)}$, $\underline{\lambda}_h = (\lambda_h^{(1)}, ..., \lambda_h^{(M)})$, $\underline{\lambda} = (\lambda^{(1)}, ..., \lambda^{(M)})$.

2. The stationary probability p_{m_1,m_2} that an arriving user (including a retrial user) sees m_l users, connected in AP l, $l = 1, 2$.

$$
p_{m_1,m_2} = \frac{1}{\Lambda} \sum_{i=1}^{M} \sum_{n=0}^{N^*} (\lambda_h^{(i)} + \lambda^{(i)} + n\alpha^{(i)}) x_{n,m_1,m_2,i}^*, \ 0 \le m_l \le M_l.
\tag{15}
$$

3. The stationary probability p_{Bf} that an arriving flex user (including a retrial flex user) is blocked.

$$
p_{Bf} = \frac{1}{\Lambda} \sum_{i=1}^{M} \sum_{m_1=L_1}^{M_1} \sum_{m_2=L_2}^{M_2} \sum_{n=0}^{\infty} (\lambda_3^{(i)} + n\alpha^{(i)}) x_{n,m_1,m_2,i}^*.
\tag{16}
$$

4. Dropping probability for handover users in AP $1, 2$

$$BP_{1h} = \sum_{m_2=0}^{M_2} \sum_{n=0}^{\infty} \sum_{i=1}^{M} x^*_{n,M_1,m_2,i},$$
$$BP_{2h} = \sum_{m_1=0}^{M_1} \sum_{n=0}^{\infty} \sum_{i=1}^{M} x^*_{n,m_1,M_2,i}. \tag{17}$$

5. Mean number of users occupied AP 1, 2.

$$E(Q_1) = \sum_{m_1=1}^{M_1} \sum_{m_2=0}^{M_2} \sum_{n=0}^{\infty} \sum_{i=1}^{M} m_1 x^*_{n,m_1,m_2,i},$$
$$E(Q_2) = \sum_{m_1=0}^{M_1} \sum_{m_2=1}^{M_2} \sum_{n=0}^{\infty} \sum_{i=1}^{M} m_2 x^*_{n,m_1,m_2,i}. \tag{18}$$

6. Mean number of blocked flex users in orbit.

$$E(Q_r) = \sum_{m_1=0}^{M_1} \sum_{m_2=0}^{M_2} \sum_{n=1}^{N^*-1} \sum_{i=1}^{M} n x^*_{n,m_1,m_2,i} + N^* \underline{x}^*_{N^*-1} R(I - R)^{-1}. \tag{19}$$

7. The blocking probability for an arriving originating subscriber of SP 1, 2.

$$BP_{1f} = \sum_{m_1=L_1}^{M_1} \sum_{m_2=0}^{M_2} \sum_{n=0}^{\infty} \sum_{i=1}^{M} x^*_{n,m_1,m_2,i},$$
$$BP_{2f} = \sum_{m_2=L_2}^{M_2} \sum_{m_1=0}^{M_1} \sum_{n=0}^{\infty} \sum_{i=1}^{M} x^*_{n,m_1,m_2,i}. \tag{20}$$

6   Numerical Results

In the following, we proceed with a scenario to illustrate the system performance. We assume that the mean service requirement of each user is $1/\mu = 50$ Mbit. Moreover, we assume that the system operates in two state random environment defined by $S = \begin{pmatrix} -2 & 2 \\ 3 & -3 \end{pmatrix}$, and its invariant vector is $\underline{s} = (0.6, 0.4)$. According to the environmental state, let also $(r_-^{(1)}, r_-^{(2)}) = (2, 4)$, $(r_+^{(1)}, r_+^{(2)}) = (5, 10)$, $(C_1^{(1)}, C_1^{(2)}) = (50, 100)$, $(C_2^{(1)}, C_2^{(2)}) = (40, 80)$ all in Mbits/s, to be the minimum/maximum attainable transfer rate, and the offered bandwidth by the APs,

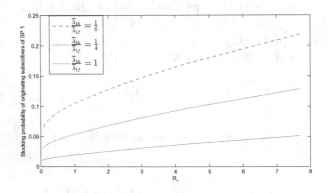

Fig. 2. BP_{1f} vs. R_1

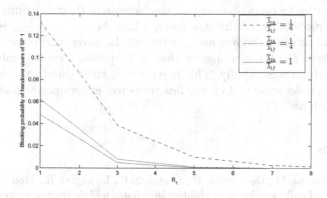

Fig. 3. BP_{1h} vs. R_1

respectively. Then, the maximum number of connected users at AP 1,2 are $M_1 = 25$, $M_2 = 20$ respectively. We reserve $R_2 = 6$ service positions only for handover users in the AP 2, and assume that the mean dwelling times in the cell, according to the environmental process, are given by $(1/\nu_1, 1/\nu_2) = (2\ min, 5\ min)$. It is further assumed that the rethinking time in order a blocked flex user to attempt a retrial is exponentially distributed with rates $(\alpha^{(1)}, \alpha^{(2)}) = (2, 1)$ subject to the environmental state.

In Fig. 2 it is seen that the dropping probability for the originating subscribers of SP 1 expectably increases when the randomized number of guard channels increases, where $\overline{\lambda}_{kh} = \underline{s}\Lambda_{kh}\underline{1}'$, $\overline{\lambda}_{kf} = \underline{s}\Lambda_{kf}\underline{1}'$ is the average arrival rate of handover and originating subscriber users respectively. Moreover, by increasing the ratio $\sigma_1 = \overline{\lambda}_{1h}/\overline{\lambda}_{1f}$ the dropping probability of originating subscribers of SP 1 decreases. On the other hand, in Fig. 3 it is seen, as expected, that the dropping probability of handover users of SP 1 decreases when we increase the reserved bandwidth. Since the smaller the σ_1, implies the reduction of the arrival rate of the handover users, we expect that the BP_{1h} will also decrease.

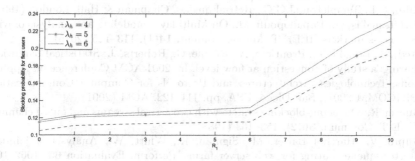

Fig. 4. p_{Bf} vs. R_1

Finally, in Fig. 4 we can observe that the increase in R_1 will definitely increase the blocking probability of the flex users, where $\lambda_h = \overline{\lambda}_{1h} + \overline{\lambda}_{2h}$, is the total average arrival rate of handover users at the cell. However, this increase is slower at lower levels of R_1. We can observe that as R_1 passes a certain value close to R_2, then p_{Bf} increases rapidly. This is expected, since due to the lack of free resources (i.e., a decrease in L_1), the flex users will experience serious problems to connect with an AP.

References

1. Ajmone Marsan, M., De Carolis, G., Leonardi, E., Lo Cigno, R., Meo, M.: Efficient estimation of call blocking probabilities in cellular mobile telephony networks with customer retrials. IEEE J. Sel. Areas in Commun. **19**, 332–346 (2001)
2. Artalejo, J.R., Gomez-Corral, A.: Retrial queueing systems: a computational approach. Springer, Berlin (2008)
3. Artalejo, J.R., Lopez-Herrero, M.J.: Cellular mobile networks with repeated calls operating in random environment. Comput. Oper. Res. **37**, 1158–1166 (2010)
4. Borst, S.: User-level performance of channel-aware scheduling algorithms in wireless data networks. IEEE ACM T. Network. **13**(3), 636–647 (2005)
5. Bright, L.W., Taylor, P.G.: Calculating the equilibrium distribution in level dependent quasi-birth-and-death processes. Stoch. Models **11**(3), 497–525 (1995)
6. Chang, C.-J., Su, T.-T., Chiang, Y.-Y.: Analysis of a cutoff priority cellular radio system with finite queueing and reneging/dropping. IEEE ACM T. Network. **2**(2), 166–175 (1994)
7. Choi, B.D., Chang, Y., Kim, B.: $MAP_1/MAP_2/M/c$ retrial queue with guard channels and it's application to cellular networks. TOP **7**(2), 231–248 (1999)
8. Choi, B.D., Chang, Y.: $MAP_1, MAP_2/M/c$ retrial queue with the retrial group of finite capacity and geometric loss. Math. Comput. Model. **30**, 99–114 (1999)
9. Do, T.V.: A new computational algorithm for retrial queues to cellular mobile systems with guard channels. Comput. Ind. Eng. **59**(4), 865–872 (2010)
10. Do, T.V.: Solution for a retrial queueing problem in cellular networks with the fractional guard channel policy. Math. Comput. Model. **53**(11–12), 2058–2065 (2011)
11. Economou, A., Lopez-Herrero, M.J.: Performance analysis of a cellular mobile network with retrials and guard channels using waiting and first passage time measures. Eur. T. Telecommun. **20**(14), 389–401 (2009)
12. Falin, G.I., Templeton, J.G.C.: Retrial queues. Chapman & Hall, London (1997)
13. Fortetsanakis, G., Papadopouli, M.: On Multi-layer modeling and analysis of wireless access markets. IEEE T. Mobile Comput. **14**(1), 113–125 (2015)
14. Fredj, S., Bonald, T., Proutiere, A., Regnie, G., Roberts, J.: Statistical bandwidth sharing: a study of congestion at flow level. In: 2001 ACM Conference on Applications, Technologies, Architectures, and Protocols for Computer Communications (SIGCOMM 2001), San Diego, USA, pp. 111–122. ACM (2001)
15. Guerin, R.: Queueing-blocking system with two arrival streams and guard channels. IEEE T. Commun. **36**(2), 153–163 (1988)
16. Gupta, V., Harchol Balter, M., Sigman, K., Whitt, W.: Analysis of join-the-shortest-queue routing for web server farms. Perform. Evaluation **64**, 1062–1081 (2007)

17. Hong, D.H., Rappaport, S.S.: Traffic model and performance analysis for cellular mobile radio telephone systems with prioritized and non-prioritized handoff procedures. IEEE Trans. Veh. Technol. **VT–28**, 77–92 (1986)
18. Kim, C.S., Klimenok, V.I., Lee, S.C., Dudin, A.N.: The BMAP/PH/1 retrial queueing system operating in random environment. J. Stat. Plan. Infer. **137**, 3904–3916 (2007)
19. Kim, C.S., Klimenok, V., Musho, V., Dudin, A.: The BMAP/PH/N retrial queueing system operating in Markovian random environment. Comput. Oper. Res. **37**, 1228–1237 (2010)
20. Kim, C.S., Klimenok, V.I., Dudin, A.N.: Analysis and optimization of guard channel policy in cellular mobile networks with account of retrials. Comput. Oper. Res. **43**, 181–190 (2014)
21. Klessig, H., Fehske, A., Fettweis, G.: Admission control in interference-coupled wireless data networks: a queuing theory-based network model. In: 12th International Symposium on Modeling and Optimization in Mobile, Ad Hoc, and Wireless Networks (WiOpt 2014), pp. 151–158. IEEE Press. doi:10.1109/WIOPT.2014.6850293
22. Knessl, C.: On finite capacity processor-sharing queues. SIAM J. Appl. Math. **50**(1), 264–287 (1990)
23. Latouche, G., Ramaswami, R.: Introduction to matrix analytic methods in stochastic modeling. ASA-SIAM, Philadelphia (1999)
24. Liu, X., Fapojuwo, A.: Performance analysis of hierarchical cellular networks with queueing and user retrials. Int. J. Commun. Syst. **19**, 699–721 (2006)
25. Litjens, R., Roijers, F., van den Berg, J., Boucherie, R., Fleuren, M.: Performance analysis of wireless LANs: an integrated packet/flow-level approach. In: 18th International Teletraffic Congress - ITC-18, Berlin, pp. 931–940 (2003)
26. Machihara, F., Saitoh, M.: Mobile customers model with retrials. Eur. J. Oper. Res. **189**, 1073–1087 (2008)
27. Mancuso, V., Alouf, S.: Reducing costs and pollution in cellular networks. IEEE Commun. Mag. **49**, 63–71 (2011)
28. Neuts, M.F.: Matrix-geometric solutions in stochastic models. The John Hopkins University Press, Baltimore (1981)
29. Neuts, M.F., Rao, B.M.: Numerical investigation of a multiserver retrial model. Queueing Syst. **7**, 169–190 (1990)
30. Nunez-Queija, R.: Sojourn times in non-homogeneous QBD processes with Processor-Sharing. Stoch. Models **17**(1), 61–92 (2001)
31. Nunez-Queija, R., van den Berg, J.L., Mandjes, M.R.H.: Performance evaluation of strategies for integration of elastic and stream traffic. In: Smith, D., Key, P. (eds.) 16th International Teletraffic Congress - ITC-16, pp. 1039–1050. Elsevier, Amsterdam (1999)
32. Phung-Duc, T., Masuyama, H., Kasahara, S., Takahashi, Y.: A simple algorithm for the rate matrices of level-dependent QBD processes. In: 5th International Conference on Queueing Theory and Network Applications, pp. 46–52 (2010)
33. Pla, V., Casares-Giner, V.: Analysis of priority channel assignment schemes in mobile cellular communication systems: a spectral theory approach. Perform. Evaluation **59**, 199–224 (2005)
34. Ramaswami, V., Taylor, P.G.: Some properties of the rate operators in level dependent quasi-birth-and-death processes with countable number of phases. Stoch. Models **12**, 143–164 (1996)

35. Roszik, J., Sztrik, J., Virtamo, J.: Performance analysis of finite-source retrial queues operating in random environments. Int. J. Oper. Res. **2**, 254–268 (2007)
36. Song, W., Zhuang, W.: Interworking of Wireless LANs and Cellular Networks. Springer, New York (2012)
37. Tran-Gia, P., Mandjes, M.: Modeling of customer retrial phenomenon in cellular mobile networks. IEEE J. Sel. Areas Commun. **15**, 1406–1414 (1997)
38. Wu, Y., Williamson, C., Luo, J.: On processor sharing and its applications to cellular data network provisioning. Perform. Evaluation **64**, 892–908 (2007)

Efficient Performance Evaluation of Wireless Networks with Varying Channel Conditions

Ekaterina Evdokimova$^{(\boxtimes)}$, Koen De Turck, Sabine Wittevrongel, and Dieter Fiems

Department of Telecommunications and Information Processing, Ghent University, St-Pietersnieuwstraat 41, 9000 Gent, Belgium
{ekaterina.evdokimova,koen.deturck,sabine.wittevrongel,dieter.fiems}
@telin.ugent.be

Abstract. This paper investigates the performance of opportunistic schedulers in wireless networks. A base station communicates over fading channels with multiple mobile nodes, each experiencing varying and not necessarily identical wireless channel conditions. An opportunistic scheduler optimises performance by accounting for both buffer size as well as channel conditions when allocating the transmitter energy among its users. The present study provides the necessary analytical tools to assess performance of opportunistic schedulers both fast and accurately, thereby allowing for fast evaluation and comparison of scheduling algorithms. The scheduler is modelled as a Markovian queueing system with multiple finite queues in a random environment. Already for a limited number of users and limited buffer capacities, the size of the state space of the Markov model makes the direct calculation of the steady-state probability vector nearly impossible. Therefore, we rely on Maclaurin series expansions so as to study the scheduler under light traffic conditions as well as in overload. The computational complexity for calculating the first N terms in the series expansions is $O(NM^2S)$, where M is the size of the state space of the exogenous channel process and S is the size of the state space of the entire Markov chain.

Keywords: Performance analysis · Opportunistic scheduling · Wireless networks · Markov analysis · Steady-state distribution

1 Introduction

Efficiently allocating networking resources is key for the performance of many multi-user (MU) communication systems. In wireline communications, such allocation aims at optimising performance metrics like network throughput, delay and jitter, while at the same time retaining fairness between its users [1]. Wireline allocation however does not directly apply to MU wireless communication systems, most prominently due to bandwidth limitations and due to the time-variation of the channel conditions of its mobile users [2].

Time-variability of the channel conditions can be exploited by opportunistically transmitting to good channels. Opportunistic scheduling is a promising

© Springer International Publishing Switzerland 2015
M. Gribaudo et al. (Eds.): ASMTA 2015, LNCS 9081, pp. 59–72, 2015.
DOI: 10.1007/978-3-319-18579-8_5

cross-layer method that holds the potential of significantly improving wireless networks' efficiency in the near future. Such scheduling however immediately introduces the trade-off between wireless efficiency — a preference to schedule to the best channel — and fair scheduling — each user is entitled to a certain amount of network resources. Since the introduction of opportunistic scheduling in [3], numerous schedulers have been proposed for different instances of wireless networks, such as mobile cellular networks, cognitive radio, MIMO systems, see [2,4–10] and the references therein.

The present paper proposes a mathematical framework for studying opportunistic scheduling. Specifically, we propose tools for fast performance evaluation of wireless networks equipped with one access point (AP) serving multiple mobile users under varying transmission conditions. All users feedback their channel conditions, so the full channel state information (CSI) is assumed to be known by the AP. Only few authors assess performance of opportunistic schedulers by analytic means, most assessments of schedulers relying on simulations [11–13]. This is not surprising as stochastic models of opportunistic schedulers involve multiple queues. This results in Markov models with multidimensional state space. Even for a limited number of buffers (or mobile nodes) and limited buffer capacities, the state space of the Markov chain is huge which makes direct solution techniques numerically infeasible.

In [11], the problem of analytical performance evaluation was tackled nevertheless, by means of a decomposition method. The approach relies on representing the MU system with K users as a deterministic and stochastic Petri net (DSPN) with a decomposition into K subnets. Since the subnets can be analysed separately, the MU system can be represented with far fewer states than the original Markov model, and thus a low computational complexity can be achieved. A drawback of this approach is that it rules out most interactions between the queues and these are essential for a complete performance study. Indeed, the interaction is key for the scheduler as each allocation decision impacts all queues. The authors make a set of assumptions, which may lead to significant errors while calculating performance characteristics. A similar decomposition approach is presented in [12] in the context of cognitive radio spectrum allocation. Here, a queueing model is analysed by using matrix-analytical methods. However, the study is mainly focused on the single-queue case with an extrapolation to multiple queues.

In contrast to the references above, our approach studies the Markov model as is, without such approximate decomposition. In particular, we consider a queueing model with K queues, each queue corresponding to the buffer at the AP of a particular mobile node. The transmission environment is characterised by an exogenous Markov process with a finite number of states in accordance with [14]. As the size of the state space of the overall Markov process makes a direct solution technique computationally infeasible, we rely on Taylor series expansions [15–18] to assess the performance fast and accurately. Depending on their application, series expansion techniques for Markov chains are referred to as perturbation techniques, the power series method or light-traffic analysis.

While the naming is not absolute, perturbation methods are mainly motivated by the assessment of the sensitivity of the performance measures with respect to a system parameter. The case where the perturbation does not preserve the class-structure of the non-perturbed chain — the so-called singular perturbations — has received much attention in literature [15,19]. The power series method transforms a Markov chain of interest in a set of Markov chains parametrised by a possibly artificial parameter γ. For $\gamma = 0$, the chain is not only easily solved, but one can also obtain the series expansion in γ. For $\gamma = 1$ one gets the original Markov chain such that the series expansion can be used to approximate the solution of the original Markov chain, provided the convergence region of the series expansion includes $\gamma = 1$ [20]. Finally, light-traffic analysis often corresponds to the series expansion in the arrival rate at a queue. For an overview on the technique of series expansions in stochastic systems, we further refer the reader to the surveys in [21] and [22] and the recent book [23]. The present study most closely relates to the numerical series expansion approach of [17] and [18]. In contrast to this work, the present unperturbed chain is not upper-diagonal, but block upper-diagonal. It is shown below, that calculating the terms in the series expansion — in overload as well as under light traffic — is much easier than solving the queueing model for any particular load. Previous queueing-theoretic studies on multi-user multi-packet transmission systems include [13,24], but do not consider the inherent multidimensionality of the queueing problem. In order to validate the proposed performance evaluation method we validate the accuracy of our results by simulation.

The remainder of the paper is organised as follows. The next section introduces the modelling assumptions and settles the notational conventions. The proposed analysis technique is then outlined in Section III and numerically validated in Section IV. Finally, conclusions are drawn in Section V.

2 Opportunistic Scheduling Model

In this paper, we assess the performance of a wireless AP serving multiple mobile nodes under varying transmission channel conditions and opportunistic scheduling. To this end, we consider a Markovian queueing model with K finite-capacity queues that share a common transmission channel (see Figure 1). Each queue corresponds to the AP buffer of a particular mobile node. Let C_k be the capacity of the kth queue. In every queue, the arrival process is assumed to be a Poisson process; let λ_k denote the arrival rate in queue k. As the different queues share a common transmission channel, the service rates in the queues are coupled.

We make the following assumptions.

- There is an exogenous continuous-time Markov process $M(t)$ that modulates the state of the wireless transmission channel. Let $\mathcal{M} = \{1, 2, \ldots, M\}$ be the state space of this Markovian background process and let α_{ij} denote the transition rate from state i to state j, $i \neq j$, $i, j \in \mathcal{M}$.
- For every background state $m \in \mathcal{M}$, let $g_m = [g_{m1}, \ldots, g_{mK}]$ be a vector whose elements quantify the channel conditions as seen by the different

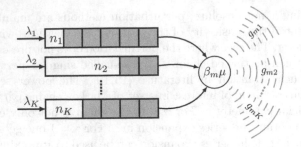

Fig. 1. Queueing model for the opportunistic scheduler

queues and let β_m be a constant denoting the overall channel quality. Let n_k be the number of packets in the kth queue and let $\mathbf{n} = [n_1, \ldots, n_K]$. Given $\mathbf{n}$, the service rate $\mu_k(\mathbf{n}, m)$ of the kth queue while $M(t) = m$ takes either one of the following forms,

$$\mu_k(\mathbf{n}, m) = \mu\beta_m \frac{g_{mk}n_k}{\sum_{\ell=1}^K g_{m\ell}n_\ell}, \tag{1}$$

$$\mu_k(\mathbf{n}, m) = \mu\beta_m \frac{g_{mk}\mathbf{1}_{\{n_k>0\}}}{\sum_{\ell=1}^K g_{m\ell}\mathbf{1}_{\{n_\ell>0\}}}, \tag{2}$$

where $\mathbf{1}_{\{X\}}$ denotes the indicator function of the event X. Note that the former service rates are inspired by discriminatory (DPS) and generalised processor sharing (GPS), respectively, but that the weights are now set to reflect the channel conditions. The exact form of these service rates is not important for the analysis though. The methodology developed further on applies to any queue-size and channel-condition dependent service rate, as long as there is channel capacity allocated to every non-empty queue in some channel state.

For further use, we introduce some additional notation. In the remainder, let $\mathcal{C}_k = \{0, 1, \ldots, C_k\}$ be the set of possible queue contents of the kth queue and let $\mathcal{C} = \mathcal{C}_1 \times \ldots \times \mathcal{C}_K$. The state space of our Markovian queueing model is then $\mathcal{C} \times \mathcal{M}$; S denotes the size of the state space $\mathcal{C} \times \mathcal{M}$. Also, $\mathbf{c} = [C_1, \ldots, C_K]$ corresponds to the case where all buffers are full; $\mathcal{M}_{\mathbf{c}} = \{[\mathbf{c}, j], j \in \mathcal{M}\}$ denotes the corresponding subset of the state space. We define $\mathbf{e}_k$ as the row vector of length K with its kth element set to 1 and all other elements zero and $\mathbf{e}$ is a row vector of ones.

At the level of abstraction of the queueing model at hand, we did not specify any technological assumptions on the AP under consideration. The model at hand allows to assess the performance of the buffer behaviour at the AP for wireless systems with opportunistic scheduling like cognitive radio, microcell networks, Wi-Fi or WiMAX networks, and for different configurations of MU MIMO with a single AP. The modelling assumptions are sufficiently versatile to capture a variety of channel- and buffer-aware policies that base their scheduling decisions on the current state of the system and transmission environment.

3 Performance Analysis

Having specified the modelling assumptions, we now present the numerical analysis technique. We first introduce the balance equations of the Markov chain under consideration. Expanding the stationary distribution of the Markov chain around $\mu = 0$ and $\lambda = 0$ yields approximations for the stationary distribution and various performance measures in the light-traffic and overload regime respectively.

3.1 Balance Equation

In view of the modelling assumptions introduced above, the state of the MU system is described by the vector $(\mathbf{n}, j)$ where $\mathbf{n} = [n_1, \ldots, n_K]$ with n_k the number of packets in the kth queue and $j \in \mathcal{M}$. Moreover, let $\pi(\mathbf{n}, j)$ be the steady-state probability to be in state $(\mathbf{n}, j)$. As there are neither simultaneous arrivals nor departures, we get the following set of balance equations:

$$\pi(\mathbf{n}, j) \left(\sum_{k=1}^{K} \left(\lambda_k \mathbf{1}_{\{n_k < C_k\}} + \mu_k(\mathbf{n}, j) \mathbf{1}_{\{n_k > 0\}} \right) + \sum_{i=1, i \neq j}^{M} \alpha_{ji} \right)$$

$$= \sum_{k=1}^{K} \pi(\mathbf{n} + \mathbf{e}_k, j) \mu_k(\mathbf{n} + \mathbf{e}_k, j) \mathbf{1}_{\{n_k < C_k\}}$$

$$+ \sum_{k=1}^{K} \pi(\mathbf{n} - \mathbf{e}_k, j) \lambda_k \mathbf{1}_{\{n_k > 0\}} + \sum_{i=1, i \neq j}^{M} \pi(\mathbf{n}, i) \alpha_{ij}, \quad (3)$$

for $\mathbf{n} \in \mathcal{C}$ and $j \in \mathcal{M}$. For ease of notation, let $\boldsymbol{\pi}(\mathbf{n}) = [\pi(\mathbf{n}, 1), \ldots, \pi(\mathbf{n}, M)]$, then we get the equivalent set:

$$\boldsymbol{\pi}(\mathbf{n}) \left(\sum_{k=1}^{K} \left(\lambda_k \mathbf{1}_{\{n_k < C_k\}} I_M + M_k(\mathbf{n}) \mathbf{1}_{\{n_k > 0\}} \right) - A \right)$$

$$= \sum_{k=1}^{K} \boldsymbol{\pi}(\mathbf{n} + \mathbf{e}_k) M_k(\mathbf{n} + \mathbf{e}_k) \mathbf{1}_{\{n_k < C_k\}}$$

$$+ \sum_{k=1}^{K} \boldsymbol{\pi}(\mathbf{n} - \mathbf{e}_k) \lambda_k \mathbf{1}_{\{n_k > 0\}}, \quad (4)$$

with $M_k(\mathbf{n})$ the $M \times M$ diagonal matrix with diagonal elements $\mu_k(\mathbf{n}, j)$, with I_M the $M \times M$ identity matrix and with A the generator matrix of $M(t)$.

3.2 Regular Perturbation

In the following subsections it is shown that a series expansion approach allows for evaluating the performance of a wireless MU system under either light-traffic

or overload conditions. In particular, it is shown that the series expansion of the
stationary solution of the Markov process corresponds to a regular perturbation
[16, 18, 25] and that the computational complexity of calculating the consecutive
terms in the series expansion is far better then the complexity of calculating
the stationary distribution directly. Prior to introducing the equations for the
system at hand, we outline the main ideas of the methodology.

The system of equations (3) takes the generic form

$$\pi Q = 0,\tag{5}$$

where π is a vector which collects all stationary probabilities $\pi(\mathbf{n}, j)$ and where
Q is a known generator matrix whose off-diagonal elements are the transition
rates between states. The row sums of the generator matrix are zero, and the
matrix has negative diagonal elements and non-negative off-diagonal elements.
Assume now that the entries of the generator matrix are affine functions of a
system parameter ϵ. In the following sections, this parameter will be the arrival
rate λ for the light-traffic approximation and the service rate μ for the overload
approximation. As the entries of the generator matrix are affine functions of ϵ,
the generic equation (5) can be written as

$$\pi^{(\epsilon)} Q = \pi^{(\epsilon)} \left(Q^{(0)} + \epsilon Q^{(1)} \right) = 0.\tag{6}$$

Here we have made the dependence of the stationary solution π on ϵ explicit.
Moreover, note that $Q^{(0)}$ is a proper generator matrix: this is the generator
matrix of the system for $\epsilon = 0$. Now, assume that this Markov process is a
uni-chain (the Markov process has at most one ergodic class). In this case,
$\pi^{(0)} Q^{(0)} = 0$ has a unique normalised solution. Moreover, by Cramer's rule, one
easily finds that $\pi^{(\epsilon)}$ is an analytic function of ϵ in an open interval around $\epsilon = 0$.
Therefore, let π_n be the nth term in the series expansion of $\pi^{(\epsilon)}$,

$$\pi^{(\epsilon)} = \sum_{n=0}^{\infty} \pi_n \epsilon^n.\tag{7}$$

Plugging the series expansion (7) in (6) and identifying equal powers of ϵ, we get

$$\pi_0 Q^{(0)} = 0, \quad \pi_{n+1} Q^{(0)} = -\pi_n Q^{(1)}.\tag{8}$$

Complementing the former set of equations with the normalisation condition,

$$\pi_0 \mathbf{e}' = 1, \quad \pi_n \mathbf{e}' = 0,\tag{9}$$

for $n > 0$ allows for recursively calculating the terms of the series expansion.

For a generic matrix $Q^{(0)}$, there is no gain in computational complexity as
one still needs to invert this matrix while solving for the next term in the series
expansion. However, for the MU system at hand, $Q^{(0)}$ has additional structure.
Indeed, for the light-traffic approximation, non-λ transitions are either departures or changes of the channel state. Assuming a proper ordering of the states

of the Markov process, the generator matrix $Q^{(0)}$ is block upper-diagonal, the blocks being the size of state space of the channel. For the overload approximation, non-μ transitions are either arrivals or changes of the channel state and — with a proper ordering of the state space — a similar block upper-diagonal structure is obtained. In either case, recursively solving the systems of equations (9) is considerably less involved. The equation

$$\pi_0 Q^{(0)} = 0$$

reduces to a system of M equations of M unknowns, while for each n the unknowns in the system

$$\pi_{n+1} Q^{(0)} = -\pi_n Q^{(1)}$$

can be solved in blocks of M unknowns at a time due to the block upper-diagonal structure of $Q^{(0)}$.

3.3   Overload-Traffic Analysis

We first consider the balance equation for $\mu \to 0$. In particular we consider the following Maclaurin series expansion of the steady-state probabilities:

$$\pi(\mathbf{n}) = \sum_{i=0}^{\infty} \pi_i(\mathbf{n}) \, \mu^i . \tag{10}$$

For ease of notation, let $\tilde{M}_k(\mathbf{n}) = \mu^{-1} M_k(\mathbf{n})$. Note that $\tilde{M}_k(\mathbf{n})$ does not depend on μ; see equations (1) and (2). Plugging the former expression into equation (4) and comparing terms in μ^i, we get

$$\pi_i(\mathbf{n}) \sum_{k=1}^{K} \lambda_k \, \mathbf{1}_{\{n_k < C_k\}} - \pi_i(\mathbf{n}) A$$

$$= \mathbf{1}_{\{i>0\}} \sum_{k=1}^{K} \pi_{i-1}(\mathbf{n} + \mathbf{e}_k) \tilde{M}_k(\mathbf{n} + \mathbf{e}_k) \, \mathbf{1}_{\{n_k < C_k\}}$$

$$- \mathbf{1}_{\{i>0\}} \pi_{i-1}(\mathbf{n}) \sum_{k=1}^{K} \tilde{M}_k(\mathbf{n}) \, \mathbf{1}_{\{n_k > 0\}}$$

$$+ \sum_{k=1}^{K} \pi_i(\mathbf{n} - \mathbf{e}_k) \lambda_k \, \mathbf{1}_{\{n_k > 0\}} . \tag{11}$$

Plugging $\mathbf{n} = \mathbf{0} = [0, 0, \ldots, 0]$ and $i = 0$ in the former equation and post-multiplying with $\mathbf{e}$ leads to

$$\pi_0(\mathbf{0}) \mathbf{e}' = 0 , \tag{12}$$

which implies $\pi_0(\mathbf{0}) = \mathbf{0}$ as the elements of $\pi_0(\mathbf{0})$ are non-negative. Using the same arguments, one then shows by iteration that for all $\mathbf{n} \in \mathcal{C} \setminus \{\mathbf{c}\}$, we have $\pi_0(\mathbf{n}) = \mathbf{0}$ and $\pi_0(\mathbf{c}) A = \mathbf{0}$. Together with the normalisation condition

$\sum_{\mathbf{n} \in \mathcal{C}} \boldsymbol{\pi}_0(\mathbf{n}) \mathbf{e}' = 1$, this shows that $\boldsymbol{\pi}_0(\mathbf{c}) = \mathbf{a}$, the steady-state solution of the Markov process $M(t)$.

For the higher-order terms $(i > 0)$, we have

$$\boldsymbol{\pi}_i(\mathbf{n}) \left(\sum_{k=1}^{K} \lambda_k \mathbf{1}_{\{n_k < C_k\}} I_M - A \right)$$

$$= \sum_{k=1}^{K} \boldsymbol{\pi}_{i-1}(\mathbf{n} + \mathbf{e}_k) \tilde{M}_k(\mathbf{n} + \mathbf{e}_k) \mathbf{1}_{\{n_k < C_k\}}$$

$$+ \sum_{k=1}^{K} \left(\boldsymbol{\pi}_i(\mathbf{n} - \mathbf{e}_k) \lambda_k - \boldsymbol{\pi}_{i-1}(\mathbf{n}) \tilde{M}_k(\mathbf{n}) \right) \mathbf{1}_{\{n_k > 0\}} . \quad (13)$$

For $\mathbf{n} \neq \mathbf{c}$, the matrix on the left-hand side is invertible. Hence, we can calculate the probabilities $\boldsymbol{\pi}_i(\mathbf{n})$ in lexicographical order. For $\mathbf{n} = \mathbf{c}$, we get

$$\boldsymbol{\pi}_i(\mathbf{c}) A = \sum_{k=1}^{K} \left(-\boldsymbol{\pi}_i(\mathbf{c} - \mathbf{e}_k) \lambda_k + \boldsymbol{\pi}_{i-1}(\mathbf{c}) \tilde{M}_k(\mathbf{c}) \right) , \quad (14)$$

and the matrix on the left-hand side is not invertible. A solution of this equation takes the form

$$\boldsymbol{\pi}_i(\mathbf{c}) = \sum_{k=1}^{K} \left(-\boldsymbol{\pi}_i(\mathbf{c} - \mathbf{e}_k) \lambda_k + \boldsymbol{\pi}_{i-1}(\mathbf{c}) \tilde{M}_k(\mathbf{c}) \right) A^{\#} + \kappa_i \mathbf{a} , \quad (15)$$

for any κ_i. Here, $A^{\#} = (A + \mathbf{e}\mathbf{a})^{-1} - \mathbf{e}\mathbf{a}$ is the group inverse of A. Finally, the remaining unknown κ_i follows from the normalisation condition

$$\sum_{\mathbf{n} \in \mathcal{C}} \boldsymbol{\pi}_i(\mathbf{n}) \mathbf{e}' = 0 . \quad (16)$$

In view of the calculations above, one easily verifies that the numerical complexity of the algorithm is $O(NM^2S)$ as there are S/M blocks, N terms in the recursion and the operations with blocks have complexity $O(M^3)$.

3.4   Light-Traffic Analysis

Similar arguments can be developed for the case of light-traffic conditions, that is, we set $\lambda_k = \lambda \tilde{\lambda}_k$ and consider an expansion of the form

$$\boldsymbol{\pi}(\mathbf{n}) = \sum_{i=0}^{\infty} \boldsymbol{\pi}_i(\mathbf{n}) \lambda^i .$$

In view of the balance equation, the terms of this series expansion adhere

$$\boldsymbol{\pi}_i(\mathbf{n}) \left(\sum_{k=1}^{K} M_k(\mathbf{n}) \, \mathbf{1}_{\{n_k > 0\}} - A \right)$$

$$= \sum_{k=1}^{K} \boldsymbol{\pi}_i(\mathbf{n} + \mathbf{e}_k) M_k(\mathbf{n} + \mathbf{e}_k) \, \mathbf{1}_{\{n_k < C_k\}}$$

$$- \mathbf{1}_{\{i>0\}} \boldsymbol{\pi}_{i-1}(\mathbf{n}) \sum_{k=1}^{K} \tilde{\lambda}_k \, \mathbf{1}_{\{n_k < C_k\}}$$

$$+ \mathbf{1}_{\{i>0\}} \sum_{k=1}^{K} \boldsymbol{\pi}_{i-1}(\mathbf{n} - \mathbf{e}_k) \tilde{\lambda}_k \, \mathbf{1}_{\{n_k > 0\}} . \quad (17)$$

For $i = 0$, we can show that $\boldsymbol{\pi}_0(\mathbf{n}) = \mathbf{0}$ for $\mathbf{n} \neq \mathbf{0}$ and $\boldsymbol{\pi}_0(\mathbf{0}) = \mathbf{a}$. For $i > 0$ and $\mathbf{n} \neq \mathbf{0}$, we can recursively calculate all $\boldsymbol{\pi}_i(\mathbf{n})$ in reverse lexicographical order as the matrix on the left-hand side is invertible. For $\mathbf{n} = \mathbf{0}$, we get

$$\boldsymbol{\pi}_i(\mathbf{0}) = (-\sum_{k=1}^{K} \boldsymbol{\pi}_i(\mathbf{e}_k) M_k(\mathbf{e}_k) + \boldsymbol{\pi}_{i-1}(\mathbf{0}) \sum_{k=1}^{K} \tilde{\lambda}_k) A^{\#} + \tilde{\kappa}_i \mathbf{a} , \quad (18)$$

where $\tilde{\kappa}_i$ can be determined from the normalisation condition (16).

4   Numerical Results

We now illustrate our approach by means of some numerical results and assess the accuracy of our approximation by means of simulation. Note that the time to simulate the Markov process (by means of the Gillespie algorithm [26]) is far longer than the time needed to calculate the approximations.

Limiting the number of parameters of the channel, we assume that the channel qualities at the receivers are varying independently and that the characteristics of all channels are identical. We therefore first focus on the Markov process which characterises a single channel. Let H be the number of states of the Markov process $M_k(t)$ for the kth channel and let $\mathbf{g} = [g_1, \ldots, g_H]$ and $\tilde{A}$ be the vector of values that quantify the quality of the channel in the different channel states and the generator matrix of the Markov process $M_k(t)$, respectively. Value g_1 corresponds to poor channel quality with no transmission, value g_H corresponds to excellent channel conditions where transmission is possible at the highest rate. The values in between represent the attenuation level of the channel capacity depending on the SNR. In order to obtain reasonable values for $\mathbf{g}$ as well as for $\tilde{A}$ in terms of channel properties, we rely on [14]. This paper however assumes that channel changes are at discrete time instants. To obtain a corresponding continuous-time Markov model, we introduce an additional rate γ and assume that the channel changes in accordance to [14] on the events of a Poisson process with rate γ.

Due to the independence of the channels, the simultaneous evolution of all channels is a Markov process as well with H^K states and with generator matrix

$A = \oplus_{n=1}^{K} \tilde{A}$. Here $\oplus$ denotes the Kronecker sum. Moreover, each state of $M(t)$ maps on a vector of states of the channel processes $M_k(t)$, such that one easily identifies g_{mk}, given the values g_k for a single channel. In addition, the parameter β_m describing the global channel condition is assumed to be equal to the average channel condition,

$$\beta_m = \frac{1}{K} \sum_{k=1}^{K} g_{mk} .$$

We consider the case of a wireless AP serving $K = 3$ mobile nodes, each having a buffer with finite capacity $C = 35$. The channel to each of the K users switches between $H = 3$ possible states of the finite-state Markov chain model with corresponding qualities $g_{mk} \in [0.3, 0.7, 1]$. Thus, the total number of background states is $M = H^K = 27$ and the overall state space size is $M(C+1)^K = 1.259.712$. The generator matrix of the background process A is obtained by means of [14] applying the following parameter values. The Doppler spread is 100 Hz and the average SNR is 20 dB, whereas the vector of SNR state thresholds is $[10, 20, 30]$dB. Assuming $\gamma = 100$ we obtain the following continuous-time process,

$$\tilde{A} = \begin{pmatrix} -0,0054 & 0,0054 & 0 \\ 0,0081 & -0,0128 & 0,0047 \\ 0 & 0,0081 & -0,0081 \end{pmatrix} .$$

Figure 2 shows the mean total queue content at the AP under light traffic (a) and for overload (b) for various orders N of the series expansions and for the opportunistic DPS policy. To verify our approximations, we also display simulation results. For sufficiently high-order expansions we observe that the approximations are accurate in the regions $\lambda = 0 \ldots 0.4$ and $\mu = 0 \ldots 1$ for light traffic and overload, respectively.

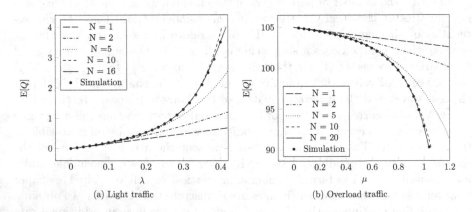

(a) Light traffic   (b) Overload traffic

Fig. 2. Mean queue content at the AP: approximation by a series expansion around $\lambda = 0$ (a) and $\mu = 0$ (b)

In order to define the region where the approximation is sufficiently accurate we rely on the simple heuristic by analogy with [18]. Let $f_N(x)$ be the Nth order expansion in x ($x = \mu$ or $x = \lambda$), then for given ε we can find a range Δx where the following condition is satisfied

$$\frac{|f_{N+1}(x) - f_N(x)|}{f_{N+1}(x)} < \varepsilon,$$

or equivalently

$$1 - \varepsilon < \frac{f_N(x)}{f_{N+1}(x)} < 1 + \varepsilon.$$

For the system under consideration and for given $\varepsilon = 0.01$, we have the regions $\Delta\lambda = 0.35$ and $\Delta\mu = 0.8$ for the light-traffic and overload cases, respectively.

For the sake of comparison, we provide both the approximations and simulations for discriminatory processor sharing (DPS) (1) and generalised processor sharing (GPS) (2) schedulers in Figure 3. As the figure demonstrates, for the light traffic case, the mean AP buffer content is almost identical under both schedulers. The reason is that the queues contain but a few packets at most, such that the difference between DPS and GPS is not outspoken. Under overload traffic on the contrary, a GPS policy outperforms DPS. Since GPS is a purely opportunistic scheduler, it will optimise throughput which results in a reduction of the total queue content. DPS also prioritises longer queues which allows for a reduction in delay jitter but comes at the price of higher mean queue content and delay (by Little's theorem).

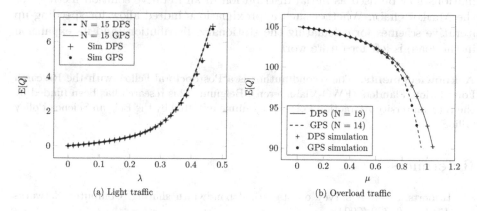

(a) Light traffic   (b) Overload traffic

Fig. 3. Mean queue content at the AP: comparison of DPS and GPS under light (a) and overload (b) traffic

5   Conclusions and Future Work

In this paper we considered a queueing model for the performance evaluation of a wireless access point under varying channel conditions. In order to represent the transmission channel variations we introduced a Markovian exogenous channel process. To solve the entire Markov chain and cope with the prohibitively large size of the state space that is inherent to queueing systems with multiple finite-capacity buffers, we relied on series expansion techniques. We showed that in contrast to the stationary solution of the Markov chain, the terms of expansion of this solution can be calculated fast and accurately, both for expansions in overload $\mu = 0$ and light traffic $\lambda = 0$. The computational complexity of the suggested method is $O(NM^2S)$, with N the order of the expansions, M the number of channel states and S the size of the state space of the overall Markov chain. Furthermore, to evaluate the region where the approximation is sufficiently accurate we applied a simple heuristic which compares the $(N+1)$th and Nth order expansions. We illustrated the proposed approach with some numerical experimentations involving the DPS and GPS scheduling policies. The approximations were found to be accurate for light traffic till reasonable load and for reasonable load till complete overload.

The present study did not address how to assess system performance for medium load. The upper- or lower-diagonal block structures for the unperturbed Markov chains in light- and overload traffic, are key for the evaluation methodology proposed, and are not present when expanding in medium load, say for λ/μ around 0.5. The light- and overload traffic expansion however do yield crude approximations for the stationary distribution in medium load. Such approximations may be used as initial distribution in an iterative solution method for the Markov chain. Whether such approximations indeed allow for speeding up iterative schemes for calculating the stationary distribution (and its expansion in the load) is left for future work.

Acknowledgments. The second author is a Postdoctoral Fellow with the Research Foundation, Flanders (FWO-Vlaanderen), Belgium. This research has been funded by the Interuniversity Attraction Poles Programme initiated by the Belgian Science Policy Office.

References

1. Roberts, J.W.: A survey on statistical bandwidth sharing. Computer Networks **45**(3), 319–332 (2004)
2. Liu, X., Chong, E.K.P., Shroff, N.B.: Opportunistic transmission scheduling with resource-sharing constraints in wireless networks. IEEE Journal on Selected Areas in Communications **19**(10), 2053–2064 (2001)
3. Knopp, R., Humblet, P.: Information capacity and power control in single-cell multiuser communications. In: Proc. IEEE ICC, pp. 331–334, June 1995
4. Liu, X., Chong, E.K.P., Shroff, N.B.: A framework for opportunistic scheduling in wireless networks. Computer Networks **41**(4), 451–474 (2003)

5. Ajib, W., Haccoun, D.: An overview of scheduling algorithms in MIMO-based fourth-generation wireless-systems. IEEE Network **19**(5), 43–48 (2005)
6. Asadi, A., Mancuso, V.: A survey on opportunistic scheduling in wireless communications. IEEE Communications Surveys & Tutorials **15**(4), 1671–1688 (2013)
7. Mietzner, J., Schober, R., Lampe, L., Gerstacker, W.H., Hoeher, P.A.: Multiple-antenna techniques for wireless communications-a comprehensive literature survey. IEEE Communications Surveys & Tutorials **11**(2), 87–105 (2009)
8. Shakkottai, S., Rappaport, T.S., Karlsson, P.C.: Cross-layer design for wireless networks. IEEE Communications Magazine **41**(10), 74–80 (2003)
9. Gesbert, D., Kountouris, M., Heath, R.W., Chae, C.B., Salzer, T.: Shifting the MIMO paradigm. IEEE Signal Processing Magazine **24**(5), 36–46 (2007)
10. Lin, X., Shroff, N.B., Srikant, R.: A tutorial on cross-layer optimization in wireless networks. IEEE Journal on Selected Areas in Communications **24**(8), 1452–1463 (2006)
11. Lei, L., Lin, C., Cai, J., Shen, X.: Performance analysis of wireless opportunistic schedulers using stochastic Petri nets. IEEE Transactions on Wireless Communications **8**(4), 2076–2087 (2009)
12. Rashid, M.M., Hossain, M.J., Hossain, E.H., Bhargava, V.K.: Opportunistic Spectrum Scheduling for Multiuser Cognitive Radio: A Queueing Analysis. IEEE Transactions on Wireless Communications **8**(10), 5259–5269 (2009)
13. Bellalta, B., Faridi, A., Barcelo, J., Daza, V., Oliver, M.: Queueing analysis in multiuser multi-packet transmission systems using spatial multiplexing. arXiv preprint arXiv:1207.3506 (2012)
14. Wang, H.S., Moayeri, N.: Finite-state Markov channel-a useful model for radio communication channels. IEEE Transactions on Vehicular Technology **44**(1), 163–171 (1995)
15. Altman, E., Avrachenkov, K.E., Núñez-Queija, R.: Perturbation analysis for denumerable markov chains with application to queueing models. Advances in Applied Probability **36**(3), 839–853 (2004)
16. De Turck, K., De Cuypere, E., Wittevrongel, S., Fiems, D.: Algorithmic approach to series expansions around transient markov chains with applications to paired queuing systems. In: Proc. of the 6th International Conference on Performance Evaluation Methodologies and Tools, pp. 38–44 (2012)
17. De Turck, K., De Cuypere, E., Fiems, D.: A Maclaurin-series expansion approach to multiple paired queues. Operations Research Letters **42**(3), 203–207 (2014)
18. De Turck, K., Fiems, D.: A series expansion approach for finite-capacity processor sharing queues. In: Proc. of the 7th International Conference on Performance Evaluation Methodologies and Tools, pp. 118–125 (2013)
19. Lasserre, J.B.: A Formula for Singular Perturbations of Markov Chains. Journal of Applied Probability **31**(3), 829–833 (1994)
20. van den Hout, W.B.: The power-series algorithm. Ph.D Thesis. University of Tilburg (1996)
21. Błaszczyszyn, B., Rolski, T., Schmidt, V.: Light-traffic approximations in queues and related stochastic models. In: Advances in Queueing: Theory, Methods and Open Problems. CRC Press, Boca Raton (1995)
22. Kovalenko, I.: Rare events in queueing theory. A survey. Queueing Systems **16**(1), 1–49 (1994)
23. Avrachenkov, K., Filar, J.A.: Analytic Perturbation Theory and Its Applications. SIAM (2014)

24. Rashid, M.M., Hossain, E., Bhargava, V.K.: Cross-layer analysis of downlink v-blast MIMO transmission exploiting multiuser diversity. IEEE Transactions on Wireless Communications 8(9), 4568–4579 (2009)
25. Avrachenkov, K.E., Haviv, M.: Perturbation of null spaces with application to the eigenvalue problem and generalized inverses. Linear Algebra and its Applications 369, 1–25 (2003)
26. Banks, H.T., Broido, A., Gayvert, K., Hu, S., Joyner, M., Link, K.: Simulation Algorithms for Continuous Time Markov Chain Models. Applied Eletromagnetics and Mechanics 37 (2007)

Mixed Networks with Multiple Classes of Customers and Restart

Jean-Michel Fourneau[1]($\boxtimes$) and Katinka Wolter[2]

[1] PRiSM, Univ. Versailles St Quentin, UMR CNRS 8144, Versailles, France
jmf@prism.uvsq.fr
[2] Institute of Computer Science, Free University Berlin, Berlin, Germany
katinka.wolter@fu-berlin.de

Abstract. We consider a network of queues with multiples classes of customers and restart signals. A restart signal makes a customer in the queue change its class and restart to the first step of service. The queues which receive signals can have an infinite server or a processor sharing discipline. The service time distributions are hyper-exponential, which can have high variance and are realistic for many real-world applications, such as transmission times over the internet. For distributions with large variability the restart mechanism can be useful. We prove that, under ergodicity condition, such a model has a product form steady-state distribution. This model contains two original features which were previously not allowed in a network of queues with negative customers: a part of the network is closed and some stations are Infinite Server queues.

1 Introduction

This paper generalizes in many directions the result obtained in [FWR+13] where G-networks with restart signals were introduced and have been shown to have a product form steady-state distribution. This analytical solution is used in a solver [WRD15]. First, we consider that the restart signal comes from an open subnetwork with positive and negative customers and triggers while in our previous work we assumed that the restart signals come from outside the network and the network only contained positive customers. We also assume a closed topology for the sub-network to which the signals arrive. Due to this topology, we are able to consider for the first time for G-networks, stations with infinite service capacity (IS queues). Finally, for the sake of simplicity, we assume that the service times follow hyper-exponential distributions which are specific to the class of the customers and the queue. Note that in the context of restart mechanism such an assumption is not really restrictive. Indeed, such a mechanism is only useful when the service times have a high variability and it may be worth to restart a service which takes too long. With a restart, a user aborts a running job that has exceeded a given deadline and submits it again to the system. In many scenarios, such a mechanism allows to reduce the total response time for a job [vMW04, vMW06]. Here we use the same queueing representation already defined in [FWR+13]. Jobs are represented as positive

© Springer International Publishing Switzerland 2015
M. Gribaudo et al. (Eds.): ASMTA 2015, LNCS 9081, pp. 73–86, 2015.
DOI: 10.1007/978-3-319-18579-8_6

customers in the G-network of queues and restarts are modeled as signals which change the class of a customer and restart the service. A trigger signal can also move the affected job among queues.

G-networks of queues with signals have received a considerable attention since the seminal paper by Gelenbe [Gel91] in 1991 where he introduced networks with positive and negative customers. A negative customer deletes a positive customer if there is any in the queue at its arrival. Then it disappears. If the queue is empty, it also disappears immediately. A negative customer is never kept in the queue. It is a signal which deletes a customer. Such a network with positive and negative customers are associated with models of Random Neural networks [Gel94] and are therefore suitable to model control algorithms.

The model we have presented in [FWR$^+$13] uses the signal approach to represent a restart in a network of queues with customers which have iid PH distributed service times. At the reception of a signal, a customer chosen according to the queuing discipline begins a restart procedure. This procedure has three steps.

1. First, it can accept or refuse the signal with a probability which depends on its class and on its step of service in the PH.
2. Second, it changes to another class using a stochastic mutation matrix.
3. Finally, it jumps to the first step of the PH distribution where it is starting again its service.

We use the same model (except the distribution of service which is simpler) to represent the effect of a restart signal but we add several features in the network concerning the topology, the service disciplines, the existence of other signals in the network (negative customers as in [Gel91] or triggers as in [Gel93]). This paper is merely theoretical as we prove that the queueing network has a product form steady-state distribution if the associated Markov chain is ergodic. The proof is based on the resolution of the global balance equation. One may use other theoretical approaches to establish the result. But it not clear that the CAT and RCAT theorems proved by Harrison [Har03, Har04, BHM10] are easier for networks with multiple classes of customer.

The result rises many interesting theoretical questions we address at the end of the paper concerning a possible mean value analysis approach for the closed sub-network with restart and triggers signals.

The technical part of the paper is organized as follows. In the next section we present the model and the proof is given in Section 3. Many details of the proof are postponed into an appendix for the sake of readability.

2   Description of the Model

We investigate generalized networks with an arbitrary number N of queues, K classes of positive customers, three types of signals (negative customers, restarts and triggers) and a mixed topology. The mixed topology is described as follows (see Fig. 1). The set of queues is divided into two subsets which constitute a proper partition. Let $S1$ and $S2$ these two sets. $S1$ is an open network. It receives fresh customers from the outside and some customers may leave queue i with probability d_i to the environment. But $S2$ is a closed network of queues. The novelty here is that signals can be sent from a queue in $S1$ into a queue of $S2$. No movement of customers are allowed between $S1$ to $S2$. Thus, the network is a mixed network of generalized queues.

More precisely, signals appear after a service of a customer in $S1$. At the completion of its service in queue i in $S1$, a customer may:

- route as a customer to a queue (say j) in $S1$ with probability $P_{i,j}$,
- route as a trigger signal to a queue (say j) in $S2$ with probability $T_{i,j}$,
- route as a restart signal to a queue (say j) in $S2$ with probability $R_{i,j}$,
- route as a negative customer signal to a queue (say j) in $S1$ with probability $Q_{i,j}$,
- leave the system with probability d_i.

Of course, we have the following normalization for all $i \in S1$:

$$d_i + \sum_{j \in S1} P_{i,j} + \sum_{j \in S1} Q_{i,j} + \sum_{j \in S2} R_{i,j} + \sum_{j \in S2} T_{i,j} = 1. \tag{1}$$

The behavior of customers in queues of $S2$ is much simpler: at the completion of its service in queue i in $S2$, a customer of class k may route as a customer of class l to a queue (say j) in $S2$ with probability $P_{i,j}^{(k,l)}$. It is important to remark that there are no negative customers in $S2$.

The model in Fig. 1 could for instance represent a system with load balancing in the following way: the open subnetwork $S1$ on the left is the control part of the system which sends restart signals to the closed queueing network $S2$ on the right side. We want to optimize the throughput inside the closed subsystem. The routing of the customers out of queue E depends on the class of the customers and their service times are also class dependent. The restart signal received at queue E is used to change the class of the customers. Therefore, the average service time but also the variance change when we restart a customer. And the routing is changed. All these parameters have an influence on the throughput in the closed subsystem and it gives us the opportunity to control the system with the restart signals sent from subnetwork $S1$.

The exemple in Fig. 1 is a very simple one: we only have one source or restart and one destination. Furthermore we do not consider triggers which move customers between queues of $S2$. Clearly we can model systems where the signals have a large impact on the throughput and using the product form solution we hope that we can optimize such models.

Let M be the number of customers in $S2$ irrespective of their location, class and step of service at the initial time. We will see in the following that this total number does not evolve with time.

We assume that the queues in $S1$ contains only one class of customers and that the service times are iid exponential. Therefore the queueing discipline is not relevant as we only study steady-state distribution. Fresh arrivals of customers

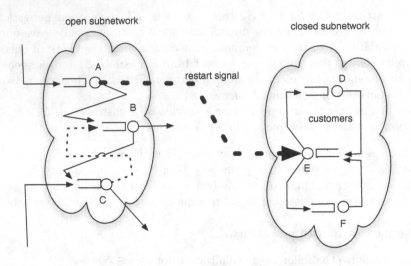

Fig. 1. Mixed Topology, the customers' movements are represented with straight lines, the negative customers by dotted lines and the restart signals by thick dotted lines

only occur at queues in $\mathcal{S}1$. The arrivals follow independent Poisson processes. The rate for arrivals at queue i in $\mathcal{S}1$ is λ_i.

The queues in $\mathcal{S}2$ contain K classes of customers. We assume that the queueing discipline is the processor sharing (PS) policy or the infinite server (IS) policy. We also assume that the service times follow hyper-exponential distributions. The hyper-exponential distribution is a mixture of H exponential distributions with probability of mixture $m_i^{k,p}$ such as:

$$\forall i, k \quad \sum_{p=1}^{H} m_i^{(k,p)} = 1. \tag{2}$$

At queue i, the intensity of service for customers of class k with index p for the mixture of distributions is denoted as $\mu_i^{(k,p)}$. Set $\mathcal{S}2$ is partitioned into two sets: $\mathcal{S}3$ and $\mathcal{S}4$. $\mathcal{S}3$ is the set of Processor Sharing queues while $\mathcal{S}4$ is the set of Infinite Server queues. It is worthy to remark that this model allows one to consider Infinite Server queues while, in general, it is not possible to find the steady-state distribution of such a queue with negative customers or more complex signals.

We represent the system by vector $\boldsymbol{x}$ whose component x_i is the state of queue i. If $i \in \mathcal{S}1$, x_i is the number of customers while if $i \in \mathcal{S}2$, x_i is the vector $(x_i^{(k,p)})$ for $k = 1, \ldots, K$ and $p = 1, \ldots, H$ where $(x_i^{(k,p)})$ is the number of customer of class k and phase p in station i. For these queues $||x_i||$ will be the number of customers in the queue.

The signals act as follows:

- a negative customer deletes a normal customer at its arrival, if there is any in the queue. Then it vanishes. If the queue is empty, the negative customer disappears immediately. A negative customer is never queued.
- a restart signal arriving at queue j selects a customer of class k and step of service p according to a distribution of probability which is denoted by $N_j^{(k,p)}$ and restarts it. The customer stays in the same queue (i.e. say j) but it changes its class (from k to l) and begins a new service according to the distribution for its new class. If it fails, the customer is kept unchanged. The matrix of the class transformation is $C_j^{(k,l)}$. The probability of the selection of a class k customer is state dependent but it also depends on the queueing discipline of the station which receives the signal: for queues in $S3$,

$$N_j^{(k,p)}(x_j) = \frac{x_j^{(k,p)}}{||x_j||}1_{||x_j||>0},$$

and for queues in $S4$,

$$N_j^{(k,p)}(x_j) = \frac{x_j^{(k,p)}}{M}1_{||x_j||>0}.$$

Of course we have for all index of class k, queue index $j \in S2$, and state x_j:

$$\sum_{l}^{K} C_j^{(k,l)} = 1, \quad and \quad \sum_{k=1}^{K}\sum_{p=1}^{H} N_j^{(k,p)}(x_j) \leq 1.$$

$1 - \sum_{k=1}^{K}\sum_{p=1}^{H} N_j^{(k,p)}(x_j)$ represents the probability that the signal fails.

- a trigger signal arriving at queue j selects a customer with the same distribution of probability than the restart and it moves it to another queue. The customer also changes its class. Let $A_{j,r}^{(k,l)}$ be the probability that the selected customer (in queue j and of class k) moves to queue r as a customer of class l. If it fails, the customer is kept unchanged. For all $j \in S2$ and class index k we have:

$$\sum_{l=1}^{K}\sum_{r\in S2} A_{j,r}^{(k,l)} = 1.$$

Clearly, $S1$ is open while $S2$ is closed for the routing of customers. Therefore we called this topology a mixed network of queues with signals. Due to the effect of the signals arriving in $S2$ which do not change the total number of customers and due to the routing of customers in $S2$, the total number of customers in $S2$ does not change. Therefore M is the total number of customers in $S2$ during the sample-path and at steady-state.

Clearly, $(x)_t$ is a Markov chain.

3   Main Result

We now state that the model introduced in the previous section has a product form steady-state distribution if the chain is ergodic and if the flow equations have a solution which satisfy the stationarity constraints. Let us first introduce a notation. In the following $S_i^{(k,p)}(x_i)$ will represent the service rate for customer of class k and step of service (also called phase) p at station i when the state is x_i. We have for all $i \in \mathcal{S}4$:

$$S_i^{(k,p)}(x_i) = \mu_i^{(k,p)} x_i^{(k,p)},$$

and for all $i \in \mathcal{S}3$,

$$S_i^{(k,p)}(x_i) = \mu_i^{(k,p)} \frac{x_i^{(k,p)}}{||x_i||} \mathbb{1}_{||x_i||>0},$$

and of course $S_i^{(k,p)}$ is zero when the queue is empty. Note that the form of N and A are essential to obtain the result. However their consistency with the service S gives a nice physical interpretation.

Theorem 1. *Consider a mixed G-network with customers and signals as described in section 2. Assume that the Markov chain is ergodic. If the system*

$$
\begin{cases}
\rho_i = \frac{\lambda_i + \sum_{j \in \mathcal{S}1} \mu_j \rho_j P_{j,i}}{\mu_i + \sum_{j \in \mathcal{S}1} \mu_j \rho_j Q_{j,i}} & \forall i \in \mathcal{S}1 \\[2mm]
a(i,k,p) = \sum_{j \in \mathcal{S}2} \sum_{l=1}^{K} \sum_{q=1}^{H} \mu_j^{(l,q)} \rho_j^{(l,q)} P_{j,i}^{(l,k)} m_i^{(k,p)} & \forall i \in \mathcal{S}2 \\[2mm]
b(i,k) = \sum_{j \in \mathcal{S}1} \sum_{l=1}^{K} \sum_{q=1}^{H} \mu_j \rho_j R_{j,i} C_i^{(l,k)} \rho_j^{(l,q)} & \forall i \in \mathcal{S}2 \\[2mm]
c(i,k) = \sum_{j \in \mathcal{S}1} \mu_j \rho_j \sum_{r \in \mathcal{S}3} T_{j,r} \sum_{q=1}^{H} \sum_{l=1}^{K} A_{r,i}^{(l,k)} \rho_r^{(l,q)} & \forall i \in \mathcal{S}2 \\[2mm]
d(i,k) = \sum_{j \in \mathcal{S}1} \mu_j \rho_j \sum_{r \in \mathcal{S}4} T_{j,r} \sum_{q=1}^{H} \sum_{l=1}^{K} A_{r,i}^{(l,k)} \rho_r^{(l,q)}/M & \forall i \in \mathcal{S}2 \\[2mm]
e(i) = \sum_{j \in \mathcal{S}1} \mu_j \rho_j (R_{j,i} + T_{j,i}) & \forall i \in \mathcal{S}2 \\[2mm]
\rho_i^{(k,p)} = m_i^{(k,p)} \frac{a(i,k,p) + b(i,k) + c(i,k) + d(i,k)}{\mu_i^{(k,p)} + e(i)} & \forall i \in \mathcal{S}3 \\[2mm]
\rho_i^{(k,p)} = m_i^{(k,p)} \frac{a(i,k,p) + b(i,k)/M + c(i,k) + d(i,k)}{\mu_i^{(k,p)} + e(i)/M} & \forall i \in \mathcal{S}4
\end{cases}
\tag{3}
$$

has a solution such that $\rho_i < 1$ for all $i \in \mathcal{S}1$ and $\sum_{k=1}^{K} \sum_{p=1}^{H} \rho_i^{(k,p)} < 1$ for $i \in \mathcal{S}2$, then the steady-state distribution has a product form distribution. More precisely,

$$\pi(x) = C \left(\prod_{i \in \mathcal{S}1} (1 - \rho_i) \rho_i^{x_i} \right) \left(\prod_{i \in \mathcal{S}2} g_i(x_i) \right),$$

where C is a normalizing constant and for all $i \in S3$,

$$g_i(x_i) = (1 - \sum_{k=1}^{K} \sum_{p=1}^{H} \rho_i^{(k,p)}) \|x_i\|! \prod_{k=1}^{K} \prod_{p=1}^{H} \frac{(\rho_i^{(k,p)})^{x_i^{(k,p)}}}{x_i^{(k,p)}!},$$

while for all $i \in S4$,

$$g_i(x_i) = \prod_{k=1}^{K} \prod_{p=1}^{H} \exp(-\rho_i^{(k,p)}) \frac{(\rho_i^{(k,p)})^{x_i^{(k,p)}}}{x_i^{(k,p)}!},$$

Proof: the proof is based on the resolution of the global balance equation. Let us first write the Kolmogorov equation at steady-state (Equation 4, next page) Let us first explain the right hand side of this equation.

- Terms [1] to [5] represent a well known G-network with negative customers as described in [Gel91].
- Term [6] models the emission of customers between queues of $S2$.
- Terms [7] and [8] represent the emission of restart signal from a queue i of $S1$ to a queue of $S2$. Term [7] describes the effect of a signal when it succeeds while Term [8] models the failure of the signal once it has been emitted.
- Terms [9] to [11] model the emission of a trigger signal initiated by a queue of $S1$ and with moves from queue j to queue r, both in $S2$. Term [9] describes the success of the signal and its effect while Terms [10] and [11] describes the failure due to empty queue (Term [10]) or due to the selection process (Term [11]).

The proof is postponed in an appendix for the sake of readability. And we just give here some simplification rules. e_i represents the state of an empty network with only one customer at queue i, $i \in S1$, while $e_i^{(k,p)}$ is the state on an empty network with only one customer of class k at queue i in $S2$ during step p of its service.

$$\frac{\pi(x + e_i)}{\pi(x)} = \rho_i \quad and \quad \frac{\pi(x + e_i - e_j)}{\pi(x)} = \frac{\rho_i}{\rho_j} \forall i, j \in S1.$$

Lemma 1. *For all $i \in S2$, the quantity $S_i^{(k,p)}(x_i + e_i^{(k,p)}) \frac{\pi(x+e_i^{(k,p)})}{\pi(x)}$ does not depend on the queuing discipline of station i.*

Proof. We compute the quantity for Processor Sharing and Infinite Server disciplines. For all $i \in S4$

$$S_i^{(k,p)}(x_i + e_i^{(k,p)}) \frac{\pi(x + e_i^{(k,p)})}{\pi(x)} = \mu_i^{(k,p)}(x_i^{(k,p)} + 1) \frac{\rho_i^{(k,p)}}{(x_i^{(k,p)} + 1)} = \mu_i^{(k,p)} \rho_i^{(k,p)}.$$

For all $i \in S3$

$$S_i^{(k,p)}(x_i + e_i^{(k,p)}) \frac{\pi(x + e_i^{(k,p)})}{\pi(x)} = \mu_i^{(k,p)} \frac{(x_i^{(k,p)} + 1)}{\|x_i\| + 1} \frac{\|x_i\| + 1}{(x_i^{(k,p)} + 1)} \rho_i^{(k,p)} = \mu_i^{(k,p)} \rho_i^{(k,p)}.$$

Thus the quantity does not depend on the queueing discipline and is equal to $\mu_i^{(k,p)} \rho_i^{(k,p)}$.

$$\pi(x)\left[\sum_{i\in S1}\mu_i 1_{x_i>0} + \sum_{i\in S1}\lambda_i + \sum_{i\in S2}\sum_{k=1}^{K}\sum_{p=1}^{H}S_i^{(k,p)}(x_i)\right] =$$

$$\sum_{i\in S1}\lambda_i\pi(x-e_i)1_{x_i>0} \qquad [1]$$

$$+\sum_{i\in S1}\mu_i\pi(x+e_i)d_i \qquad [2]$$

$$+\sum_{i\in S1}\sum_{j\in S1}\mu_i\pi(x+e_i-e_j)P_{i,j}1_{x_j>0} \qquad [3]$$

$$+\sum_{i\in S1}\sum_{j\in S1}\mu_i\pi(x+e_i+e_j)Q_{i,j} \qquad [4]$$

$$+\sum_{i\in S1}\sum_{j\in S1}\mu_i\pi(x+e_i)Q_{i,j}1_{x_j=0} \qquad [5]$$

$$+\sum_{i\in S2}\sum_{j\in S2}\sum_{k=1}^{K}\sum_{l=1}^{K}\sum_{p=1}^{H}\sum_{q=1}^{H}\pi(x+e_i^{(k,p)}-e_j^{(l,q)})S_i^{(k,p)}(x_i+e_i^{(k,p)})P_{i,j}^{(k,l)}m_j^{(l,q)}1_{x_j^{(l,q)}>0} \qquad [6]$$

$$+\sum_{i\in S1}\sum_{j\in S2}\mu_i R_{i,j}\sum_{k=1}^{K}\sum_{p=1}^{H}\sum_{l=1}^{K}\sum_{q=1}^{H}N_j^{(k,p)}(x_j+e_j^{(k,p)}-e_j^{(l,q)}) \times$$
$$C_j^{(k,l)}\pi(x+e_i+e_j^{(k,p)}-e_j^{(l,q)})m_j^{(l,q)}1_{x_j^{(l,q)}>0} \qquad [7]$$

$$+\sum_{i\in S1}\sum_{j\in S2}\mu_i R_{i,j}\pi(x+e_i)\left(1_{||x_j||=0}+1_{||x_j||>0}(1-\sum_{k=1}^{K}\sum_{p=1}^{H}N_j^{(k,p)}(x_j))\right) \qquad [8]$$

$$+\sum_{i\in S1}\sum_{j\in S2}\mu_i T_{i,j}\sum_{k=1}^{K}\sum_{p=1}^{H}N_j^{(k,p)}(x_j+e_j^{(k,p)}) \times$$
$$\sum_{l=1}^{K}\sum_{r\in S2}A_{j,r}^{(k,l)}\sum_{q=1}^{H}\pi(x+e_i+e_j^{(k,p)}-e_r^{(l,q)})m_r^{(l,q)}1_{x_r^{(l,q)}>0} \qquad [9]$$

$$+\sum_{i\in S1}\sum_{j\in S2}\mu_i T_{i,j}\pi(x+e_i)1_{||x_j||=0} \qquad [10]$$

$$+\sum_{i\in S1}\sum_{j\in S2}\mu_i T_{i,j}\pi(x+e_i)1_{||x_j||>0}(1-\sum_{k=1}^{K}\sum_{p=1}^{H}N_j^{(k,p)}(x_j)) \qquad [11]$$

$$(4)$$

Property 1. Let us now consider the product $N_j^{(k,p)}(x_j+e_j^{(k,p)}-e_j^{(l,q)})\frac{\pi(x+e_j^{(k,p)})}{\pi(x)}$ for both types of queues in $S2$. If $j\in S3$ then we get:

$$N_j^{(k,p)}(x_j+e_j^{(k,p)}-e_j^{(l,q)})\frac{\pi(x+e_j^{(k,p)})}{\pi(x)}=\frac{x_j^{(k,p)}+1}{||x_j||+1}\frac{\rho_j^{(k,p)}(||x_j||+1)}{x_j^{(k,p)}+1}=\rho_j^{(k,p)}.$$

and for queues in $\mathcal{S}4$,

$$N_j^{(k,p)}(x_j + e_j^{(k,p)} - e_j^{(l,q)})\frac{\pi(x + e_j^{(k,p)})}{\pi(x)} = \frac{x_j^{(k,p)} + 1}{M} \frac{\rho_j^{(k,p)}}{x_j^{(k,p)} + 1} = \frac{\rho_j^{(k,p)}}{M}.$$

Therefore we cannot abstract the queuing discipline when we deal with Terms [7] to [11].

Lemma 2. *The flow equation between the outside and all the queues in $\mathcal{S}1$ is:*

$$\sum_{i \in \mathcal{S}1} \lambda_i = \sum_{i \in \mathcal{S}1} \mu_i \rho_i d_i + \sum_{i \in \mathcal{S}1}\sum_{j \in \mathcal{S}1} \mu_i \rho_i \rho_j Q_{i,j} + \sum_{i \in \mathcal{S}1}\sum_{j \in \mathcal{S}1} \mu_i \rho_i Q_{i,j} + \sum_{i \in \mathcal{S}1}\sum_{j \in \mathcal{S}2} \mu_i \rho_i (R_{i,j} + T_{i,j}).$$

(5)

Remember that the network containing the queues of $\mathcal{S}2$ is closed and it does not exchange customers with the outside.

Proof. Consider the definition of ρ_i, multiply both sides by the denominator and sum for all queue in $\mathcal{S}1$.

$$\sum_{i \in \mathcal{S}1} \mu_i \rho_i + \sum_{i \in \mathcal{S}1}\sum_{j \in \mathcal{S}1} \rho_i \rho_j \mu_j Q_{j,i} = \sum_{i \in \mathcal{S}1} \lambda_i + \sum_{i \in \mathcal{S}1}\sum_{j \in \mathcal{S}1} \rho_j \mu_j P_{j,i}$$

Due to Eq. 1, we have:

$$\sum_{i \in \mathcal{S}1} P_{j,i} = 1 - d_j - \sum_{i \in \mathcal{S}1} R_{j,i} - \sum_{i \in \mathcal{S}1} T_{j,i} - \sum_{i \in \mathcal{S}1} Q_{j,i}$$

After substitution and cancellation of terms, we finally get Eq. 5.

Note that the normalization constant C is not equal to 1.0 as some states for the closed sub-network are not reachable. Thus we have to use a numerical algorithm to compute C.

4 Open Questions and Concluding Remarks

The network is decomposed into an open part and a closed part. Typically G-networks with signals have an open topology because many signals already studied in the literature imply the deletion of customers. Here if we assume that the signals are generated outside the component we study, we have a balance for the customers in queue. If the queue is empty, it remains empty. If there is a backlog, we still have the same total number of customers after the signal. Therefore it is possible to consider a closed sub-network where the total number of customers is constant. Note however that this number is not constant per class.

This opens many questions related to known results for closed queueing networks with customers and without signals. We just want to mention here the PASTA property [HP93] to describe the state seen by an arriving customer or a signal and the possibility to derive a MVA algorithm [BGDMT98] to compute the average queue size and the average delay without computing the distribution. It is well-known that in a closed Gordon Newell network, an arriving customer sees the steady-state distribution of the network with one customer less. Such a question can now be considered for closed sub-network of generalized queues with restart and trigger signals. Such a theoretical result allows to compute the average queue size without computing the steady-state distribution. Note that even if Theorem 1 states that the steady-state has a product form, we have to compute the normalizing constant by a numerical algorithm which still has to be developed. A natural idea consists in a generalization of the convolution algorithm proposed by Buzen [BGDMT98], to networks of queues with signals.

Acknowledgments. This research is partly supported by a PHC Procope grant. The French team is supported by grant ANR-12-MONU-00019.

References

[BGDMT98] Bolch, G., Greiner, S., De Meer, H., Trivedi, K.: Queueing Networks and Markov Chains. John Wiley & Sons (1998)

[BHM10] Balsamo, S., Harrison, P.G., Marin, A.: A unifying approach to product-forms in networks with finite capacity constraints. In: Misra, V., Barford, P., Squillante, M.S. (eds.) Proceedings of the 2010 ACM SIGMETRICS International Conference on Measurement and Modeling of Computer Systems, SIGMETRICS 2010, New York, pp. 25–36. ACM (2010)

[FWR+13] Fourneau, J.-M., Wolter, K., Reinecke, P., Krauß, T., Danilkina, A.: Multiple class G-networks with restart. In: ACM/SPEC International Conference on Performance Engineering, ICPE 2013, pp. 39–50. ACM (2013)

[Gel91] Gelenbe, E.: Product-form queuing networks with negative and positive customers. Journal of Applied Probability **28**, 656–663 (1991)

[Gel93] Gelenbe, E.: G-networks with instantaneous customer movement. Journal of Applied Probability **30**(3), 742–748 (1993)

[Gel94] Gelenbe, E.: G-networks: An unifying model for queuing networks and neural networks. Annals of Operations Research **48**(1–4), 433–461 (1994)

[Har03] Harrison, P.G.: Turning back time in Markovian process algebra. Theoretical Computer Science **290**(3), 1947–1986 (2003)

[Har04] Harrison, P.G.: Compositional reversed Markov processes, with applications to G-networks. Perform. Eval. **57**(3), 379–408 (2004)

[HP93] Harrison, P.G., Patel, N.M.: Performance Modelling of Communication Networks and Computer Architectures. Addison-Wesley, January 1993

[vMW04] van Moorsel, A.P.A., Wolter, K.: Analysis and algorithms for restart. In: 1st International Conference on Quantitative Evaluation of Systems (QEST 2004), The Netherlands, pp. 195–204. IEEE Computer Society (2004)

[vMW06] van Moorsel, A.P.A., Wolter, K.: Analysis of restart mechanisms in software systems. IEEE Trans. Software Eng. **32**(8), 547–558 (2006)

[WRD15] Wolter, K., Reinecke, P., Dräger, M.: GRnet: a tool for G-networks with restart. In: Proceedings of the 6th ACM/SPEC International Conference on Performance Engineering, Austin, USA, pp. 101–102. ACM (2015)

Appendix

We divide both sides by π and we use the simplification rules already mentioned.

$$\sum_{i \in S1} \mu_i 1_{x_i > 0} + \sum_{i \in S1} \lambda_i + \sum_{i \in S2} \sum_{k=1}^{K} \sum_{p=1}^{H} S_i^{(k,p)}(x_i) = \sum_{i \in S1} \lambda_i \frac{1}{\rho_i} 1_{x_i > 0} \qquad [1]$$

$$+ \sum_{i \in S1} \mu_i \rho_i d_i + \sum_{i \in S1} \sum_{j \in S1} \mu_i \frac{\rho_i}{\rho_j} P_{i,j} 1_{x_j > 0} + \sum_{i \in S1} \sum_{j \in S1} \mu_i \rho_i \rho_j Q_{i,j} + \sum_{i \in S1} \sum_{j \in S1} \mu_i \rho_i Q_{i,j} 1_{x_j = 0} \quad [2-5]$$

$$+ \sum_{i \in S2} \sum_{j \in S2} \sum_{k=1}^{K} \sum_{l=1}^{K} \sum_{p=1}^{H} \sum_{q=1}^{H} \mu_i^{(k,p)} \rho_i^{(k,p)} P_{i,j}^{(k,l)} m_j^{(l,q)} \frac{g_j(x_j - e_j^{(l,q)})}{g_j(x_j)} 1_{x_j^{(l,q)} > 0} \qquad [6]$$

$$+ \sum_{i \in S1} \sum_{j \in S2} \mu_i R_{i,j} \sum_{k=1}^{K} \sum_{p=1}^{H} \sum_{l=1}^{K} \sum_{q=1}^{H} N_j^{(k,p)}(x_j + e_j^{(k,p)} - e_j^{(l,q)}) \times$$

$$C_j^{(k,l)} \rho_i \frac{g_j(x_j + e_j^{(k,p)} - e_j^{(l,q)})}{g_j(x_j)} m_j^{(l,q)} 1_{x_j^{(l,q)} > 0} \qquad [7]$$

$$+ \sum_{i \in S1} \sum_{j \in S2} \mu_i R_{i,j} \rho_i \left(1_{||x_j||=0} + 1_{||x_j||>0} (1 - \sum_{k=1}^{K} \sum_{p=1}^{H} N_j^{(k,p)}(x_j)) \right) \qquad [8]$$

$$+ \sum_{i \in S1} \sum_{j \in S2} \mu_i T_{i,j} \sum_{k=1}^{K} \sum_{p=1}^{H} N_j^{(k,p)}(x_j + e_j^{(k,p)}) \times$$

$$\sum_{l=1}^{K} \sum_{r \in S2} A_{j,r}^{(k,l)} \sum_{q=1}^{H} \rho_i \frac{g_j(x_j + e_j^{(k,p)})}{g_j(x_j)} \frac{g_r(x_r - e_r^{(l,q)})}{g_r(x_r)} m_r^{(l,q)} 1_{x_r^{(l,q)} > 0} \qquad [9]$$

$$+ \sum_{i \in S1} \sum_{j \in S2} \mu_i T_{i,j} \rho_i 1_{||x_j||=0} \qquad [10]$$

$$+ \sum_{i \in S1} \sum_{j \in S2} \mu_i T_{i,j} \rho_i 1_{||x_j||>0} (1 - \sum_{k=1}^{K} \sum_{p=1}^{H} N_j^{(k,p)}(x_j)) \qquad [11]$$

We remark that $1_{x_j = 0} = 1 - 1_{x_j > 0}$ and that $1_{||x_j||=0} + 1_{||x_j||>0} = 1$. We substitute in Term [5] and Term [8]. We factorize Term [10] and Term [11]. We move the negative parts we have made appeared with this substitutions on the left hand side of the equation. We factorize Term [1] and Term [3] after exchanging the index in Term [3].

$$\sum_{i \in S1} \mu_i 1_{x_i > 0} + \sum_{i \in S1} \lambda_i + \sum_{i \in S2} \sum_{k=1}^{K} \sum_{p=1}^{H} S_i^{(k,p)}(x_i) + \sum_{i \in S1} \sum_{j \in S1} \mu_i \rho_i Q_{i,j} 1_{x_j > 0}$$

$$+ \sum_{i \in S1} \sum_{j \in S2} \mu_i \rho_i (R_{i,j} + T_{i,j}) 1_{||x_j||>0} \sum_{k=1}^{K} \sum_{p=1}^{H} N_j^{(k,p)}(x_j)$$

$$= \sum_{i \in S1} \frac{1}{\rho_i} 1_{x_i > 0} (\lambda_i + \sum_{j \in S1} \mu_j \rho_j P_{j,i}) \qquad [1]$$

$$+ \sum_{i \in S1} \mu_i \rho_i d_i + \sum_{i \in S1} \sum_{j \in S1} \mu_i \rho_i \rho_j Q_{i,j} + \sum_{i \in S1} \sum_{j \in S1} \mu_i \rho_i Q_{i,j} \qquad [2+3+4]$$

$$+ \sum_{i \in S2} \sum_{j \in S2} \sum_{k=1}^{K} \sum_{l=1}^{K} \sum_{p=1}^{H} \sum_{q=1}^{H} \mu_i^{(k,p)} \rho_i^{(k,p)} P_{i,j}^{(k,l)} m_j^{(l,q)} \frac{g_j(x_j - e_j^{(l,q)})}{g_j(x_j)} 1_{x_j^{(l,q)} > 0} \qquad [5]$$

$$+ \sum_{i \in S1} \sum_{j \in S2} \mu_i R_{i,j} \sum_{k=1}^{K} \sum_{p=1}^{H} \sum_{l=1}^{K} \sum_{q=1}^{H} N_j^{(k,p)}(x_j + e_j^{(k,p)} - e_j^{(l,q)}) \times$$

$$C_j^{(k,l)} \rho_i \frac{g_j(x_j + e_j^{(k,p)} - e_j^{(l,q)})}{g_j(x_j)} m_j^{(l,q)} 1_{x_j^{(l,q)} > 0} \quad [6]$$

$$+ \sum_{i \in S1} \sum_{j \in S2} \mu_i \rho_i (R_{i,j} + T_{i,j}) \quad [7]$$

$$+ \sum_{i \in S1} \sum_{j \in S2} \mu_i T_{i,j} \sum_{k=1}^{K} \sum_{p=1}^{H} N_j^{(k,p)}(x_j + e_j^{(k,p)}) \times$$

$$\sum_{l=1}^{K} \sum_{r \in S2} A_{j,r}^{(k,l)} \sum_{q=1}^{H} \rho_i \frac{g_j(x_j + e_j^{(k,p)})}{g_j(x_j)} \frac{g_r(x_r - e_r^{(l,q)})}{g_r(x_r)} m_r^{(l,q)} 1_{x_r^{(l,q)} > 0} \quad [8]$$

Due to the definition of ρ_i for $i \in S1$, Term [1] of the r.h.s cancels with the first and the fourth term of the l.h.s. And applying Lemma 2, the second term of the l.h.s cancels with Terms [2], [3], [4], and [7]. We get:

$$\sum_{j \in S2} \sum_{k=1}^{K} \sum_{p=1}^{H} S_j^{(k,p)}(x_j) + \sum_{j \in S2} \sum_{i \in S1} \mu_i \rho_i (R_{i,j} + T_{i,j}) 1_{\|x_j\| > 0} \sum_{k=1}^{K} \sum_{p=1}^{H} N_j^{(k,p)}(x_j)$$

$$= \sum_{j \in S2} \sum_{i \in S2} \sum_{k=1}^{K} \sum_{l=1}^{K} \sum_{p=1}^{H} \sum_{q=1}^{H} \mu_i^{(k,p)} \rho_i^{(k,p)} P_{i,j}^{(k,l)} m_j^{(l,q)} \frac{g_j(x_j - e_j^{(l,q)})}{g_j(x_j)} 1_{x_j^{(l,q)} > 0} \quad [1]$$

$$+ \sum_{j \in S2} \sum_{i \in S1} \mu_i R_{i,j} \sum_{k=1}^{K} \sum_{p=1}^{H} \sum_{l=1}^{K} \sum_{q=1}^{H} N_j^{(k,p)}(x_j + e_j^{(k,p)} - e_j^{(l,q)}) \times$$

$$C_j^{(k,l)} \rho_i \frac{g_j(x_j + e_j^{(k,p)} - e_j^{(l,q)})}{g_j(x_j)} m_j^{(l,q)} 1_{x_j^{(l,q)} > 0} \quad [2]$$

$$+ \sum_{j \in S2} \sum_{i \in S1} \mu_i T_{i,j} \sum_{k=1}^{K} \sum_{p=1}^{H} N_j^{(k,p)}(x_j + e_j^{(k,p)}) \times$$

$$\sum_{l=1}^{K} \sum_{r \in S2} A_{j,r}^{(k,l)} \sum_{q=1}^{H} \rho_i \frac{g_j(x_j + e_j^{(k,p)})}{g_j(x_j)} \frac{g_r(x_r - e_r^{(l,q)})}{g_r(x_r)} m_r^{(l,q)} 1_{x_r^{(l,q)} > 0} \quad [3]$$

We now have to consider two cases to substitute the values of $S_j^{(k,p)}$ and $N_j^{(k,p)}$ according to the queueing discipline. If queue j is in $S4$, we get:

$$\sum_{j \in S4} \sum_{k=1}^{K} \sum_{p=1}^{H} x_j^{(k,p)} \left(\mu_j^{(k,p)} + \sum_{i \in S1} \mu_i \rho_i (R_{i,j} + T_{i,j}) \frac{1}{M} \right)$$

$$= \sum_{j \in S4} \sum_{i \in S2} \sum_{k=1}^{K} \sum_{l=1}^{K} \sum_{p=1}^{H} \sum_{q=1}^{H} \mu_i^{(k,p)} \rho_i^{(k,p)} P_{i,j}^{(k,l)} m_j^{(l,q)} \frac{x_j^{(l,q)}}{\rho_j^{(l,q)}} \quad [1]$$

$$+ \sum_{j \in S4} \sum_{i \in S1} \mu_i \rho_i R_{i,j} \sum_{k=1}^{K} \sum_{p=1}^{H} \sum_{l=1}^{K} \sum_{q=1}^{H} \frac{\rho_j^{(k,p)}}{M} C_j^{(k,l)} \frac{x_j^{(l,q)}}{\rho_j^{(l,q)}} m_j^{(l,q)} \quad [2]$$

$$+ \sum_{r \in S4} \sum_{j \in S2} \sum_{i \in S1} \mu_i \rho_i T_{i,j} \sum_{k=1}^{K} \sum_{p=1}^{H} \sum_{l=1}^{K} \sum_{q=1}^{H} A_{j,r}^{(k,l)} \frac{x_r^{(l,q)}}{\rho_r^{(l,q)}} m_r^{(l,q)} \frac{g_j(x_j + e_j^{(k,p)})}{g_j(x_j)} N_j^{(k,p)}(x_j + e_j^{(k,p)}) \quad [3]$$

Note that the step functions are useless and they have been removed. According to Prop. 1, one must decompose the Term [3] into two terms based on the type of queues to be able to simplify the expression. Finally, after reordering to make it clearer:

$$\sum_{j \in S4} \sum_{k=1}^{K} \sum_{p=1}^{H} x_j^{(k,p)} \left(\mu_j^{(k,p)} + \sum_{i \in S1} \mu_i \rho_i (R_{i,j} + T_{i,j}) \frac{1}{M} \right)$$

$$= \sum_{j \in S4} \sum_{l=1}^{K} \sum_{q=1}^{H} \frac{x_j^{(l,q)}}{\rho_j^{(l,q)}} m_j^{(l,q)} \sum_{i \in S2} \sum_{k=1}^{K} \sum_{p=1}^{H} \mu_i^{(k,p)} \rho_i^{(k,p)} P_{i,j}^{(k,l)} \qquad [1]$$

$$+ \sum_{j \in S4} \sum_{l=1}^{K} \sum_{q=1}^{H} \frac{x_j^{(l,q)}}{\rho_j^{(l,q)}} m_j^{(l,q)} \sum_{i \in S1} \sum_{k=1}^{K} \sum_{p=1}^{H} \mu_i \rho_i R_{i,j} C_j^{(k,l)} \frac{\rho_j^{(k,p)}}{M} \qquad [2]$$

$$+ \sum_{r \in S4} \sum_{l=1}^{K} \sum_{q=1}^{H} \frac{x_r^{(l,q)}}{\rho_r^{(l,q)}} m_r^{(l,q)} \sum_{j \in S3} \sum_{i \in S1} \mu_i \rho_i T_{i,j} \sum_{k=1}^{K} \sum_{p=1}^{H} A_{j,r}^{(k,l)} \rho_j^{(k,p)} \qquad [3]$$

$$+ \sum_{r \in S4} \sum_{l=1}^{K} \sum_{q=1}^{H} \frac{x_r^{(l,q)}}{\rho_r^{(l,q)}} m_r^{(l,q)} \sum_{j \in S4} \sum_{i \in S1} \mu_i \rho_i T_{i,j} \sum_{k=1}^{K} \sum_{p=1}^{H} A_{j,r}^{(k,l)} \frac{\rho_j^{(k,p)}}{M} \qquad [4]$$

This relation holds because of the form of $\rho_i^{(k,p)}$ for Infinite Service queues. Let us now consider queues in $S3$ (i.e. queues with Processor Sharing discipline).

$$\sum_{j \in S3} \sum_{k=1}^{K} \sum_{p=1}^{H} \frac{x_j^{(k,p)}}{||x_j||} \left(\mu_j^{(k,p)} + \sum_{i \in S1} \mu_i \rho_i (R_{i,j} + T_{i,j}) \right)$$

$$= \sum_{j \in S3} \sum_{i \in S2} \sum_{k=1}^{K} \sum_{l=1}^{K} \sum_{p=1}^{H} \sum_{q=1}^{H} \mu_i^{(k,p)} \rho_i^{(k,p)} P_{i,j}^{(k,l)} m_j^{(l,q)} \frac{x_j^{(l,q)}}{\rho_j^{(l,q)} ||x_j||} \qquad [1]$$

$$+ \sum_{j \in S3} \sum_{i \in S1} \mu_i \rho_i R_{i,j} \sum_{k=1}^{K} \sum_{p=1}^{H} \sum_{l=1}^{K} \sum_{q=1}^{H} \rho_j^{(k,p)} C_j^{(k,l)} \frac{x_j^{(l,q)}}{\rho_j^{(l,q)} ||x_j||} m_j^{(l,q)} \qquad [2]$$

$$\qquad\qquad\qquad\qquad (6)$$

$$+ \sum_{r \in S3} \sum_{j \in S2} \sum_{i \in S1} \mu_i \rho_i T_{i,j} \sum_{k=1}^{K} \sum_{p=1}^{H} \sum_{l=1}^{K} \sum_{q=1}^{H} A_{j,r}^{(k,l)} \frac{x_r^{(l,q)}}{\rho_r^{(l,q)} ||x_r||} m_r^{(l,q)} \times$$

$$\frac{g_j(x_j + e_j^{(k,p)})}{g_j(x_j)} N_j^{(k,p)}(x_j + e_j^{(k,p)}) \qquad [3]$$

We decompose again the third term in the l.h.s according to the type of queue which receives the trigger to simplify the ratio of probabilities. After reordering we get:

$$\sum_{j \in S3} \sum_{k=1}^{K} \sum_{p=1}^{H} \frac{x_j^{(k,p)}}{||x_j||} \left(\mu_j^{(k,p)} + \sum_{i \in S1} \mu_i \rho_i (R_{i,j} + T_{i,j}) \right)$$

$$= \sum_{j \in S3} \sum_{l=1}^{K} \sum_{q=1}^{H} m_j^{(l,q)} \frac{x_j^{(l,q)}}{\rho_j^{(l,q)} ||x_j||} \sum_{i \in S2} \sum_{k=1}^{K} \sum_{p=1}^{H} \mu_i^{(k,p)} \rho_i^{(k,p)} P_{i,j}^{(k,l)} \qquad [1]$$

$$+ \sum_{j \in S3} \sum_{l=1}^{K} \sum_{q=1}^{H} \frac{x_j^{(l,q)}}{\rho_j^{(l,q)} ||x_j||} m_j^{(l,q)} \sum_{i \in S1} \mu_i \rho_i R_{i,j} \sum_{k=1}^{K} \sum_{p=1}^{H} \rho_j^{(k,p)} C_j^{(k,l)} \qquad [2] \qquad (7)$$

$$+ \sum_{r \in S3} \sum_{l=1}^{K} \sum_{q=1}^{H} \frac{x_r^{(l,q)}}{\rho_r^{(l,q)} ||x_r||} m_r^{(l,q)} \sum_{j \in S3} \sum_{i \in S1} \mu_i \rho_i T_{i,j} \sum_{k=1}^{K} \sum_{p=1}^{H} A_{j,r}^{(k,l)} \rho_j^{(k,p)} \qquad [3]$$

$$+ \sum_{r \in S3} \sum_{l=1}^{K} \sum_{q=1}^{H} \frac{x_r^{(l,q)}}{\rho_r^{(l,q)} ||x_r||} m_r^{(l,q)} \sum_{j \in S4} \sum_{i \in S1} \mu_i \rho_i T_{i,j} \sum_{k=1}^{K} \sum_{p=1}^{H} A_{j,r}^{(k,l)} \frac{\rho_j^{(k,p)}}{M} \qquad [4]$$

This relation holds because of the definition of $\rho_j^{(k,p)}$ for an Infinite Server queue.

Interconnected Wireless Sensors with Energy Harvesting

Erol Gelenbe[1] and Andrea Marin[2]([envelope])

[1] Intelligent Systems and Networks Group,
Department of Electrical and Electronic Engineering,
Imperial College, London SW7 2BT, UK
e.gelenbe@imperial.ac.uk
[2] Dipartimento di Scienze Ambientali, Informatica e Statistica,
Università Ca' Foscari di Venezia, via Torino 155, Venezia, Italy
marin@dais.unive.it

Abstract. This paper studies interconnected wireless sensors with the paradigm of Energy Packet Networks (EPN) which were previously introduced. In the EPN model, both data transmissions and the flow of energy are discretized, so that an energy packet (EP) is the minimum amount of energy (say in microjules) that is needed to process and transmit a data packet (DP) or to process a job. Previous work has modeled such systems to determine the relation between energy flow and DP transmission, or to study the balance between energy and the processing of jobs in Cloud Servers. The lack of energy, in addition to processing times, is the main source of latency in networks of sensor nodes. Thus this paper models this phenomenon, and shows that under some reasonable conditions, assuming feedforward flow of data packets and local consumption and leakage of energy, such networks have product form solutions.

1 Introduction and Previous Work

Information and communication technologies (ICT) steadily increase their energy consumption by about 4% per year [22] reaching roughly 5% of the worldwide electrical energy consumption in 2012 [18], but there is also hope that ICT can also reduce the energy consumption in other areas such as transportation for daily commutes [17,26]. However the users of ICT use more complex multimedia technologies [14] which are ever more demanding in energy because of their computational complexity and communication bandwidth, so that progress will be needed to reduce limit this growth through more efficient microelectronics and new technologies such as energy harvesting for computation and communications [25,23,21,1,24,16]. Thus recent research has addressed new technologies based on energy harvesting and so as to minimise the non-renewable energy consumption for given communication tasks [12,13].

Furthermore, earlier work [15] has also shown that smart routing [7] based on QoS can be also used to reduce overall energy consumption in a network.

© Springer International Publishing Switzerland 2015
M. Gribaudo et al. (Eds.): ASMTA 2015, LNCS 9081, pp. 87–99, 2015.
DOI: 10.1007/978-3-319-18579-8_7

Because of the random nature of both data flows and of energy, in such contexts it become convenient to view not just data and computation but also energy itself in discrete units. This has given rise to the energy packet (EP) paradigm [9,8] based on mathematical models such as G-Networks [6]. Such discrete representations are useful also to capture the stochastic nature of compute-communications, energy harvesting and data sensing in interconnected micro-electronic and computer-communication systems.

Contribution. In this paper we pursue a modeling approach developed in [10,11] where energy harvesting wireless sensors are modeled, assuming that data collection times and the time needed to harvest significant amount of energy, is substantially higher than the time needed to transmit a packet when energy is available. Thus the stochastic system representation that is used assumes finite and positive data and energy arrival rates to nodes and zero service or data transmission times when energy is available. In this paper this approach is developed for a sensor network that contains two nodes and packets travel in feedforward mode through two nodes, or just through one node, before successfully exiting the network. The structure we consider also includes not just energy harvesting, but also the realistic situation when energy leakage may occur. We prove that, under mild conditions, the equilibrium distribution of the continuous time Markov chain underlying the model has product-form solution and hence the derivation of the performance indices can be carried out efficiently. Indeed, many models of wireless sensor networks suffer the problem of the state-space explosion (see, e.g., [20,5,4]) that makes the exact analysis of the underlying stochastic process very difficult especially for large networks. Product-form allows us to derive the performance measures of the WSN by the analysis of each sensor *as if* it were isolated. From a theoretical point of view, although the model can be seen as belonging to the wide class of G-networks [6], the product-form is new and depends on some conditions on the model's rates that will be discussed later.

Structure of the Paper. The paper is structured as follows. In Section 2 we introduce the mathematical model of a single sensor. Section 3 describes the model of interconnected sensors and proves the product-form equilibrium distribution from which we derive some mean performance indices such as the expected number of data packets enqueued in a sensor and its throughput. Finally, Section 4 concludes the paper.

2   The Mathematical Model

In this section we consider a single wireless sensor which operates with energy harvesting. We assume that as soon as the sensor has both a data packet to transmit and enough energy to transmit that packet, the transmission takes place very rapidly so that it may be represented as a "zero time" or instantaneous transmission. We denote each device by the acronym EHWS (Energy Harvesting

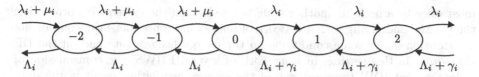

Fig. 1. CTMC underlying the model of sensor i

Wireless Sensor). The EHWS has an unlimited buffer for data packets and an unlimited "buffer" or battery for energy "packets" since we represent energy in discrete units. Thus an *energy packet* (EP) is the exact amount of energy required to transmit a data packet. Just as data packets are assumed to be collected into the EHWS in discrete packets of data, we consider that the harvested energy is also collected into the device's storage battery in discrete units (the energy packets).

Thus for an EHWS i, the state can be represented by an integer n_i, where i identifies the EHWS, where $n_i = 0$ means that the device has neither energy nor data packets, while $n_i > 0$ means that it currently stores n_i data packets but no energy, while $n_i < 0$ means that it stores n_i energy packets but no data packets. We also suppose that the EHWS harvests energy packets at a rate Λ_i while it collects data packets at a rate λ_i. Also, each device looses energy through leakage at rate μ_i, and we will assume that packets themselves will be discarded with a time-out represented by a rate γ_i.

If one considers a single EHWS whose state is represented by the integer n_1, it becomes clear that it may be modelled as a random walk, provided all the rates are parameters of independent exponentially distributed random variables, and that this random walk is ergodic provided that $\Lambda_1 + \gamma_1 > \lambda_1$ and $\lambda_1 + \mu_1 > \Lambda_1$ as in Figure 1. As a consequence of the exponential assumption, DPs and EPs arrive at the node according to independent and homogeneous Poisson processes.

Furthermore, we can readily see that its stationary distribution for an isolated node i is given by:

$$\pi_i(n_i) = \pi_i(0) \left(\frac{\lambda_i}{\Lambda_i + \gamma_i} \right)^{n_i} \quad \text{if } n_i > 0,$$

$$\pi_i(n_i) = \pi_i(0) \left(\frac{\Lambda_i}{\lambda_i + \mu_i} \right)^{-n_i} \quad \text{if } n_i < 0,$$

$$\pi_i(0) = \left[1 + \frac{\frac{\lambda_i}{\Lambda_i + \gamma_i}}{1 - \frac{\lambda_i}{\Lambda_i + \gamma_i}} + \frac{\frac{\Lambda_i}{\lambda_i + \mu_i}}{1 - \frac{\Lambda_i}{\lambda_i + \mu_i}} \right]^{-1}.$$

3   Interconnected Sensor Nodes

In this section we study the steady-state behaviour of a network of EHWSs. We assume the topology of the network to be such that each DP is forwarded at most once by a node to another node before leaving the system. We prove that the CTMC underlying such a network of sensors has a product-form equilibrium distribution under some conditions on the energy leakage rate and on the DP time-outs. In this setting, in a network of several EHWS, the transmission of a data packet (DP) from any one of the sensors may either result in the data packet arriving at the second sensor, or it may be directed towards the "exit" so that it is removed from the network. Therefore, a DP may visit at most two nodes, and that this occurs in one of four ways:

- A DP arrives from outside the network (e.g. through sensing) at one of the sensors; if that EHWS has no energy, then it is placed in the DP buffer.
- The DP arrives from outside the network at sensor i that does have energy; it consumes energy and is transmitted, then it leaves the network with probability p_{i0}.
- The DP arrives from outside the network at sensor i that does have energy; it consumes energy and is transmitted, and then arrives at sensor j with probability p_{ij}, $j \neq i$. If sensor j *does not* have energy, the DP stays there.
- Finally the DP arrives from outside the network to sensor i that does have energy; it consumes energy and is transmitted with probability p_{ij} to sensor $j \neq i$; if sensor j *does* have energy, the DP leaves the network.

A network whose topology satisfies the conditions of the two-steps routing is shown in Figure 2.

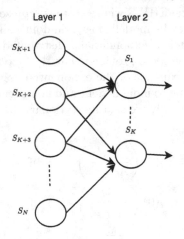

Fig. 2. A network of EHWSs with two layers. In this case packets are forwarded by at most one node.

3.1   Product-Form Analysis

In order to derive the product-form equilibrium distribution for a network of EHWSs we assume that the rate associated with the time-out (energy leakage)

Table 1. Table of notation for sensor i

Λ_i	Arrival rate of EPs at sensor i
λ_i	Arrival rate of DPs at sensor i from outside
γ_i	Time-out rate for DPs in state $n_i > 1$
γ_i^0	Time-out rate for DPs in state $n_i = 1$
μ_i	Leakage rate for EPs in state $n_i < -1$
μ_i^0	Leakage rate for EPs in state $n_i = -1$
p_{ij}	Probability of routing from EHWS i to EHWS j
$n_i > 0$	Number of DPs buffered by the sensor
$n_i < 0$	Number of EPs stored by the sensor battery

in states $n_i > 1$ ($n_i < -1$) may be different from that in state $n_i = 1$ ($n_i = -1$). We consider a network of N EHWSs in which a DP is forwarded at most once. Let us consider a network of N nodes whose state is $\mathbf{n} = (n_1, \ldots, n_N)$ with $n_i \in \mathbb{N}$. Formally, let $\mathbf{P} = (p_{ij})$ be the routing matrix with $1 \leq i, j \leq N$ and let p_{i0} be the probability that a DP leaves the network after visiting node i, i.e., $\sum_{n=1}^{N} p_{in} + p_{i0} = 1$. We recall that each sensor is described by the set of parameters shown in Table 1.

The transition rates of the CTMC underlying the network of EHWSNs are $q(\mathbf{n}, \mathbf{n}') =$

- $\Lambda_i p_{ij}$ if $\mathbf{n}' = \mathbf{n} - \mathbf{e}_i + \mathbf{e}_j$ and $n_i > 0$ (transmission of a DP from node i to j due to the harvesting of an EP)
- $\lambda_i p_{ij}$ if $\mathbf{n}' = \mathbf{n} - \mathbf{e}_i + \mathbf{e}_j$ and $n_i < 0$ (transmission of a DP from node i to node j due to the availability of a new data)
- Λ_i if $\mathbf{n}' = \mathbf{n} - \mathbf{e}_i$ and $n_i \leq 0$ (harvesting of a new EP from node i)
- $\Lambda_i p_{i0} + \gamma_i^0$ if $\mathbf{n}' = \mathbf{n} - \mathbf{e}_i$ and $n_i = 1$ (transmission of a DP to the outside or timeout of a DP at node i)
- $\Lambda_i p_{i0} + \gamma_i$ if $\mathbf{n}' = \mathbf{n} - \mathbf{e}_i$ and $n_i > 1$ (transmission of a DP to the outside or timeout of a DP at node i)
- λ_i if $\mathbf{n}' = \mathbf{n} + \mathbf{e}_i$ (generation of a new DP at node i)
- $\lambda_i p_{i0} + \mu_i^0$ if $n_i = -1$ (transmission of a DP from node i to the outside when EP are available or leakage of an EP)
- $\lambda_i p_{i0} + \mu_i$ if $n_i < -1$ (transmission of a DP from node i to the outside when EP are available or leakage of an EP)

We define for each EHWS the following quantity v_i.

$$v_i = \lambda_i + \sum_{n=1}^{N} \frac{\Lambda_n \lambda_n}{\Lambda_n + \gamma_n} p_{ni} .$$

We will show that v_i denotes the total arrival rate of DPs at EHWS i.

The following assumption on the time-out and energy leakage rates will be sufficient to prove the product-form stationary distribution of the EHWS network.

Assumption 1. We assume the following relations on the timeout settings:

- $v_i + \mu_i = \Lambda_i + \gamma_i$
- $\mu_i^0 = v_i + 2\mu_i$
- $\gamma_i^0 = \Lambda_i + 2\gamma_i$

Theorem 1. *Under the constraints of Assumption 1, given two interconnected EHWS j and k, their joint equilibrium distribution has the following product-form:*

$$\pi(n_j, n_k) = G g_j(n_j) g_k(n_k) , \tag{1}$$

where

$$g_i(n_i) = \begin{cases} 1 & \text{if } n_i = 0 \\ \frac{v_i}{\Lambda_i + \gamma_i^0} \left(\frac{v_i}{\Lambda_i + \gamma_i} \right)^{n_i - 1} & \text{if } n_i \geq 1 \\ \frac{\Lambda_i}{v_i + \mu_i^0} \left(\frac{\Lambda_i}{v_i + \mu_i} \right)^{-n_i - 1} & \text{if } n_i \leq -1 \end{cases} \tag{2}$$

with $i = j, k$ and G is the normalising constant:

$$G = \left(1 + \frac{\frac{v_j}{\Lambda_j + \gamma_j^0}}{1 - \frac{v_j}{\Lambda_j + \gamma_j}} + \frac{\frac{\Lambda_j}{v_j + \mu_j^0}}{1 - \frac{\Lambda_j}{v_j + \mu_j}} \right)^{-1} \left(1 + \frac{\frac{v_k}{\Lambda_k + \gamma_k^0}}{1 - \frac{v_k}{\Lambda_k + \gamma_k}} + \frac{\frac{\Lambda_k}{v_k + \mu_k^0}}{1 - \frac{\Lambda_k}{v_k + \mu_k}} \right)^{-1} .$$

The proof is given in appendix.

Remark 1 (On the conditions for the product-form). It is worth of notice that the product-form expression given by Theorem 1 is subject to the constraints on the rates stated in Assumption 1. The fact that some product-form results have conditions which depend on the model's transition rates is not new (see e.g., the conditions on the service rate in the First Come First Service queues of the BCMP theorem [3], or the product-forms derived in [2,19]). Nevertheless, in this case the conditions required by Assumption 1 are not strict since they give a relation on the time-out setting. Informally, we require that the sum of the energy harvesting rate and the packet time-out rate (consumption rate of packets) must be equal to the sum of the DPs arrival rate and the energy leakage rate (consumption rate of energy packets). This balance basically states that we must consume the DPs with the same rate at which we consume the EPs and can be reached by opportunely setting the time-out rate of DPs. Similar considerations hold for the time-out taking the model to state 0.

The product-form expression given by Theorem 1 can be used to compute some important performance indices such as a EHWS throughput, energy efficiency and expected number of enqueued DPs.

Proposition 1. *In stability, the DP throughput of EHWS i is:*

$$TH_i = \frac{\Lambda_i v_i}{\mu_i + v_i} . \tag{3}$$

The proof follows from the simplification of the expression:

$$\sum_{n=-\infty}^{\infty} G_i g_i(n) v_i + \sum_{n=1}^{\infty} G_i g_i(n) \Lambda_i,$$

where G_i is the normalising constant for $g_i(n)$. The total power consumption of the EHWS is clearly Λ EPs for unit of time. However, not all this energy is used to transmit DPs because some will suffer the energy leakage. The efficiency of the EHWS is given by the ratio of the EPs used for transmitting data and the total EPs consumed.

Proposition 2. *In stability, the efficiency of EHWS i is:*

$$\eta_i = 1 - \frac{\mu_i}{\gamma_i + \Lambda_i}. \tag{4}$$

Indeed, the rate of consumption of EPs for DPs transmission is given by:

$$G_i \left[\sum_{n=-\infty}^{-1} g_i(n) v_i + \sum_{n=1}^{\infty} g_i(n) \Lambda_i \right].$$

Finally, we can derive the expected number of DPs in the queue.

Proposition 3. *In stability, the expected number of DPs in the queue of EHWS i is:*

$$N_i = \frac{\gamma_i \lambda_i}{\mu_i(\gamma_i + \mu_i)}.$$

The expression of N_i can be derived by the simplification of the sum:

$$G_i \sum_{n=1}^{\infty} n g_i(n).$$

4   Conclusion

This paper has shown that a plausible and novel model of two interconnected energy harvesting wireless sensors with discretised energy harvesting and storage, and a feedforward data packet communication pattern has product form solution in its state that represents both the amount of energy and the data packet backlog at each sensor. In future work we expect that these results can be generalised to other topologies and to arbitrarily large networks.

References

1. Alippi, C., Galperti, C.: An adaptive system for optimal solar energy harvesting in wireless sensor network nodes. IEEE Transactions on Circuits and Systems I: Regular Papers **55**(6), 1742–1750 (2008)

2. Balsamo, S., Harrison, P.G., Marin, A.: Methodological construction of product-form stochastic Petri nets for performance evaluation. Journal of Systems and Software **85**(7), 1520–1539 (2012)
3. Forest Baskett, K., Chandy, M., Muntz, R.R., Palacios, F.G.: Open, closed, and mixed networks of queues with different classes of customers. J. ACM **22**(2), 248–260 (1975)
4. Bugliesi, M., Gallina, L., Hamadou, S., Marin, A., Rossi, S.: Behavioural equivalences and interference metrics for mobile ad-hoc networks. Performance Evaluation **73**, 41–72 (2014)
5. Gallina, L., Han, T., Kwiatkowska, M.Z., Marin, A., Rossi, S., Spano, A:. Automatic energy-aware performance analysis of mobile ad-hoc networks. In: Proc. of Wireless Days, pp. 1–6 (2012)
6. Gelenbe, E.: The first decade of g-networks. European Journal of Operational Research **126**(2), 231–232 (2000)
7. Gelenbe, E.: Sensible decisions based on qos. Computational Management Science **1**(1), 1–14 (2003)
8. Gelenbe, E.: Energy packet networks: adaptive energy management for the cloud. In: Proceedings of the 2nd International Workshop on Cloud Computing Platforms, p. 1. ACM (2012)
9. Gelenbe, E.: Energy packet networks: ict based energy allocation and storage. In: Rodrigues, J.J.P.C., Zhou, L., Chen, M., Kailas, A. (eds.) GreenNets 2011. LNICST, vol. 51, pp. 186–195. Springer, Heidelberg (2012)
10. Gelenbe, E.: A sensor node with energy harvesting. ACM SIGMETRICS Performance Evaluation Review **42**(2), 37–39 (2014)
11. Gelenbe, E.: Synchronising energy harvesting and data packets in a wireless sensor. Energies **8**(1), 356–369 (2015)
12. Gelenbe, E., Gesbert, D., Gündüz, D., Külah, H., Uysal-Biyikoglu, E.: Energy harvesting communication networks, optimization and demonstration: the e-crops project. In: 24th TIWDC, Tyrrhenian International Workshop 2013 on Digital Communications: Green ICT. IEEE Xplore (2013)
13. Gelenbe, E., Gündüz, D.: Optimum power level for communications with interference. In: 24th TIWDC, Tyrrhenian International Workshop 2013 on Digital Communications: Green ICT. IEEE Xplore (2013)
14. Gelenbe, E., Hussain, K., Kaptan, V.: Simulating autonomous agents in augmented reality. Journal of Systems and Software **74**(3), 255–268 (2005)
15. Gelenbe, E., Morfopoulou, C.: A framework for energy-awar routing in packet networks. Computer Journal **54**(6), 850–859 (2011)
16. Gelenbe, E., Oklander, B.: Cognitive users with useful vacations. In: 2013 IEEE International Conference on Communications Workshops (ICC), pp. 370–374. IEEE Explore (2013)
17. The Climate Group and GeSI. Smart 2020: Enabling the low carbon economy in the information age. Global E-Sustainability Initiative (2008)
18. Lannoo, B., Lambert, S., Van Heddeghem, W., Pickavet, M., Kuipers, F., Koutitas, G., Niavis, H., Satsiou, A., Beck, M.T., Fischer, A., de Meer, H., Alcock, P., Papaioannou, T., Viet, N.H., Plagemann, T., Aracil, J.: Overview of ict energy consumption (deliverable 8.1). EU Project FP7-2888021, European Network of Excellence in Internet Science, February 2013
19. Marin, A., Balsamo, S., Harrison, P.G.: Analysis of stochastic Petri nets with signals. Perf. Eval. **85**(7), 1520–1539 (2012)
20. Merro, M., Ballardin, F., Sibilio, E.: A timed calculus for wireless systems. Theoretical Computer Science **412**(47), 6585–6611 (2011)

21. Meshkati, F., Poor, H.V., Schwartz, S.C., Mandayam, N.B.: An energy-efficient approach to power control and receiver design in wireless data networks. IEEE Transactions on Communications **53**(11), 1885–1894 (2005)
22. Pettey, C.: Gartner estimates ict industry accounts for 2 percent of global co2 emissions (2007). https://www.gartner.com/newsroom/id/503867, 14:2013
23. Rodoplu, V., Meng, T.H.: Bits-per-joule capacity of energy-limited wireless networks. IEEE Transactions on Wireless Communications **6**(3), 857–865 (2007)
24. Seah, W.K.G., Eu, Z.A., Tan, H.-P.: Wireless sensor networks powered by ambient energy harvesting (wsn-heap)-survey and challenges. In: 1st International Conference on Wireless Communication, Vehicular Technology, Information Theory and Aerospace & Electronic Systems Technology, Wireless VITAE 2009, pp. 1–5. IEEE (2009)
25. Uysal-Biyikoglu, E., Prabhakar, B., El Gamal, A.: Energy-efficient packet transmission over a wireless link. IEEE/ACM Transactions on Networking (TON) **10**(4), 487–499 (2002)
26. Yu, Y., Bhatti, S.N.: The cost of virtue: reward as well as feedback are required to reduce user ict power consumption. In: Proceedings of the 5th International Conference on Future Energy Systems, pp. 157–169. ACM (2014)

A   Proof of Theorem 1

Since the network structure has a topology in which a packet if forwarded by at most one EHWS, we can prove the theorem by just considering a tandem of two sensors in the state space (n_1, n_2). Equation (1) of Theorem 1 can be rewritten as:

$$\pi(n_1, n_2) = G g_1(n_1) g_2(n_2)$$

and

$$v_1 = \lambda_1, \quad v_2 = \lambda_2 + \frac{\Lambda_1 \lambda_1 p_{12}}{\Lambda_1 + \gamma_1}.$$

For the sake of simplicity we give the proof for $p_{12} = 1$ and $\lambda_2 = 0$. The proof proceeds by substitution in the system of global balance equations of the underlying CTMC. Let us consider the case in which $n_1 > 0$. Then, the corresponding balance equation of a state (n_1, n_2), with $n_2 \subset \mathbf{Z}$ is:

$$\pi(n_1, n_2)\big(\lambda_1 + \Lambda_1 + \gamma_1^0 \delta_{n_1=1} + \gamma_1 \delta_{n_1>1} + \Lambda_2 + \gamma_2^0 \delta_{n_2=1}$$
$$+ \gamma_2 \delta_{n_2>1} + \mu_2^0 \delta_{n_2=-1} + \mu_2 \delta_{n_2<-1}\big)$$
$$= \underbrace{\pi(n_1+1, n_2)\gamma_1}_{A} + \underbrace{\pi(n_1, n_2+1)\left(\Lambda_2 + \gamma_2 \delta_{n_2>0} + \gamma_2^0 \delta_{n_2=0}\right)}_{B} + \underbrace{\pi(n_1-1, n_2)\lambda_1}_{C}$$
$$+ \underbrace{\pi(n_1+1, n_2-1)\Lambda_1}_{D} + \underbrace{\pi(n_1, n_2-1)(\mu_2 \delta_{n_2\leq-1} + \mu_2^0 \delta_{n_2=0})}_{E}$$

We divide the RHS by $\pi(n_1, n_2)$. We have:

$$\frac{A}{\pi(n_1, n_2)} = \frac{\lambda_1 \gamma_1}{\Lambda_1 + \gamma_1} \tag{5}$$

For part B:

$$\frac{B}{\pi(n_1, n_2)} = \frac{\lambda_1 \Lambda_1}{\Lambda_1 + \gamma_1} \frac{1}{\Lambda_2 + \gamma_2 \delta_{n_2 > 0} + \gamma_2^0 \delta_{n_2 = 0}}$$
$$\cdot \left(\Lambda_2 + \gamma_2 \delta_{n_2 > 0} + \gamma_2^0 \delta_{n_2 = 0} \right) \delta_{n_2 \geq 0}$$
$$+ \left(\frac{\Lambda_1 \lambda_1}{\Lambda_1 + \gamma_1} + \mu_2 \delta_{n_2 < -1} + \mu_2^0 \delta_{n_2 = -1} \right) \frac{1}{\Lambda_2} \Lambda_2 \delta_{n_2 < 0}$$
$$= \frac{\Lambda_1 \lambda_1}{\Lambda_1 + \gamma_1} + \mu_2 \delta_{n_2 < -1} + \mu_2^0 \delta_{n_2 = -1} \quad (6)$$

For part C:

$$\frac{C}{\pi(n_1, n_2)} = \frac{\Lambda_1 + \gamma_1 \delta_{n_1 > 1} + \gamma_1^0 \delta_{n_1 = 1}}{\lambda_1} \lambda_1 = \Lambda_1 + \gamma_1 \delta_{n_1 > 1} + \gamma_1^0 \delta_{n_1 = 1} \quad (7)$$

For part D:

$$\frac{D}{\pi(n_1, n_2)} = \frac{\lambda_1}{\Lambda_1 + \gamma_1} \left(\frac{\Lambda_2}{\frac{\Lambda_1 \lambda_1}{\Lambda_1 + \gamma_1} + \mu_2 \delta_{n_2 \leq -1} + \mu_2^0 \delta_{n_2 = 0}} \right) \Lambda_1 \delta_{n_2 \leq 0}$$
$$+ \frac{\lambda_1}{\Lambda_1 + \gamma_1} \left((\Lambda_2 + \gamma_2^0 \delta_{n_2 = 1} + \gamma_2 \delta_{n_2 > 1}) \frac{\Lambda_1 + \gamma_1}{\Lambda_1 \lambda_1} \right) \Lambda_1 \delta_{n_2 > 0}$$

which simplifies to:

$$\frac{D}{\pi(n_1, n_2)} = \frac{\lambda_1 \Lambda_2 \Lambda_1}{\Lambda_1 \lambda_1 + (\Lambda_1 + \gamma_1)(\mu_2 \delta_{n_2 \leq -1} + \mu_2^0 \delta_{n_2 = 0})} \delta_{n_2 \leq 0} + \Lambda_2 \delta_{n_2 > 0}$$
$$+ \gamma_2^0 \delta_{n_2 = 1} + \gamma_2 \delta_{n_2 > 1} \quad (8)$$

For part E:

$$\frac{E}{\pi(n_1, n_2)} = \frac{\Lambda_2}{\frac{\Lambda_1 \lambda_1}{\Lambda_1 + \gamma_1} + \mu_2 \delta_{n_2 \leq -1} + \mu_2^0 \delta_{n_2 = 0}} (\mu_2 \delta_{n_2 \leq -1} + \mu_0 \delta_{n_2 = 0})$$
$$= \frac{\Lambda_2 (\Lambda_1 + \gamma_1)}{\Lambda_1 \lambda_1 + (\Lambda_1 + \gamma_1)(\mu_2 \delta_{n_2 \leq -1} + \mu_2^0 \delta_{n_2 = 0})} (\mu_2 \delta_{n_2 \leq -1} + \mu_0 \delta_{n_2 = 0}) \quad (9)$$

Summing Equations (8) and (9) we have:

$$\frac{D + E}{\pi(n_1, n_2)}$$
$$= \frac{\lambda_1 \Lambda_2 \Lambda_1 + \Lambda_2 \Lambda_1 (\mu_2 \delta_{n_2 \leq -1} + \mu_2^0 \delta_{n_2 = 0}) + \Lambda_2 \gamma_1 (\mu_2 \delta_{n_2 \leq -1} + \mu_2^0 \delta_{n_2 = 0})}{\Lambda_1 \lambda_1 + (\Lambda_1 + \gamma_1)(\mu_2 \delta_{n_2 \leq -1} + \mu_2^0 \delta_{n_2 = 0})} \delta_{n_2 \leq 0}$$
$$+ \Lambda_2 \delta_{n_2 > 0} + \gamma_2^0 \delta_{n_2 = 1} + \gamma_2 \delta_{n_2 > 1}$$
$$= \frac{\lambda_1 \Lambda_2 \Lambda_1 + \Lambda_2 (\mu_2 \delta_{n_2 \leq -1} + \mu_2^0 \delta_{n_2 = 0})(\Lambda_1 + \gamma_1)}{\Lambda_1 \lambda_1 + (\Lambda_1 + \gamma_1)(\mu_2 \delta_{n_2 \leq -1} + \mu_2^0 \delta_{n_2 = 0})} \delta_{n_2 \leq 0} + \Lambda_2 \delta_{n_2 > 0} + \gamma_2^0 \delta_{n_2 = 1}$$
$$+ \gamma_2 \delta_{n_2 > 1} = \Lambda_2 + \gamma_2^0 \delta_{n_2 = 1} + \gamma_2 \delta_{n_2 > 1} \quad (10)$$

Finally, we sum Equations (5), (6), (7), (10) and obtain:

$$\frac{\lambda_1 \gamma_1}{\Lambda_1 + \gamma_1} + \frac{\Lambda_1 \lambda_1}{\Lambda_1 + \gamma_1} + \mu_2 \delta_{n_2 < -1} + \mu_2^0 \delta_{n_2 = -1}$$

$$+ \Lambda_1 + \gamma_1 \delta_{n_1 > 1} + \gamma_1^0 \delta_{n_1 = 1} + \Lambda_2 + \gamma_2^0 \delta_{n_2 = 1} + \gamma_2 \delta_{n_2 > 1}$$

$$= \lambda_1 + \mu_2 \delta_{n_2 < -1} + \mu_2^0 \delta_{n_2 = -1} + \Lambda_1 + \gamma_1 \delta_{n_1 > 1} + \gamma_1^0 \delta_{n_1 = 1} + \Lambda_2 + \gamma_2^0 \delta_{n_2 = 1} + \gamma_2 \delta_{n_2 > 1}$$

which is exactly the LHS of Equation (1) divided by $\pi(n_1, n_2)$, as required.

We now consider the case $n_1 = 0$. The balance equations for states $(n, 0)$ are:

$$\pi(0, n_2)(\lambda_1 + \Lambda_1 + \Lambda_2 + \gamma_2 \delta_{n_2 > 1} + \gamma_2^0 \delta_{n_2 = 1} + \mu_2 \delta_{n_2 < -1} + \mu_2^0 \delta_{n_2 = -1})$$

$$= \underbrace{\pi(1, n_2 - 1)\Lambda_1}_{A} + \underbrace{\pi(-1, n_2 - 1)\lambda_1}_{B} + \underbrace{\pi(-1, n_2)\mu_1^0}_{C} + \underbrace{\pi(1, n_2)\gamma_1^0}_{D}$$

$$+ \underbrace{\pi(0, n_2 + 1)\left(\Lambda_2 + \gamma_2 \delta_{n_2 \geq 1} + \gamma_2 \delta_{n_2 = 0}\right)}_{E} + \underbrace{\pi(0, n_2 - 1)\left(\mu_2 \delta_{n_2 \leq -1} + \mu_2^0 \delta_{n_2 = 0}\right)}_{F}$$

$$(11)$$

Let us compute $(A + B)/\pi(0, n_2)$:

$$\frac{\lambda_1}{\Lambda_1 + \gamma_1^0}\left(\frac{\Lambda_2}{\frac{\Lambda_1 \lambda_1}{\Lambda_1 + \gamma_1} + \mu_2^0 \delta_{n_2 = 0} + \mu_2 \delta_{n_2 < 0}}\right)\Lambda_1 \delta_{n_2 \leq 0}$$

$$+ \frac{\lambda_1}{\Lambda_1 + \gamma_1^0}\left(\frac{\Lambda_2 + \gamma_2^0 \delta_{n_2 = 1} + \gamma_2 \delta_{n_2 > 1}}{\frac{\Lambda_1 \lambda_1}{\Lambda_1 + \gamma_1}}\right)\Lambda_1 \delta_{n_2 > 0}$$

$$+ \frac{\Lambda_1}{\lambda_1 + \mu_1^0}\left(\frac{\Lambda_2}{\frac{\Lambda_1 \lambda_1}{\Lambda_1 + \gamma_1} + \mu_2^0 \delta_{n_2 = 0} + \mu_2 \delta_{n_2 < 0}}\right)\Lambda_1 \delta_{n_2 \leq 0}$$

$$+ \frac{\Lambda_1}{\lambda_1 + \mu_1^0}\left(\frac{\Lambda_2 + \gamma_2^0 \delta_{n_2 = 1} + \gamma_2 \delta_{n_2 > 1}}{\frac{\Lambda_1 \lambda_1}{\Lambda_1 + \gamma_1}}\right)\Lambda_1 \delta_{n_2 > 0}$$

$$= \left(\frac{\lambda_1 \Lambda_1}{\Lambda_1 + \gamma_1^0} + \frac{\lambda_1 \Lambda_1}{\lambda_1 + \mu_1^0}\right)\left(\frac{\Lambda_2}{\frac{\Lambda_1 \lambda_1}{\Lambda_1 + \gamma_1} + \mu_2^0 \delta_{n_2 = 0} + \mu_2 \delta_{n_2 < 0}}\delta_{n_2 \leq 0}\right.$$

$$\left. + \frac{\Lambda_2 + \gamma_2^0 \delta_{n_2 = 1} + \gamma_2 \delta_{n_2 > 1}}{\frac{\Lambda_1 \lambda_1}{\Lambda_1 + \gamma_1}}\delta_{n_2 > 0}\right)$$

Notice that by Assumption 1 we have

$$\Lambda_1 + \gamma_1^0 = 2(\Lambda_1 + \gamma_1) = 2(\lambda_1 + \mu_1) = \lambda_1 + \mu_1^0. \qquad (12)$$

This allows us to rewrite the first term of the product as $\lambda_1 \Lambda_1 / (\Lambda_1 + \gamma_1)$ and hence simplify the expression as follows:

$$\frac{A + B}{\pi(0, n_2)} = \frac{\lambda_1 \Lambda_1 \Lambda_2}{\Lambda_1 \lambda_1 + (\mu_2^0 \delta_{n_2 = 0} + \mu_2 \delta_{n_2 \leq -1})(\Lambda_1 + \gamma_1)}\delta_{n_2 \leq 0}$$

$$+ \Lambda_2 \delta_{n_2 > 0} + \gamma_2^0 \delta_{n_2 = 1} + \gamma_2 \delta_{n_2 > 1} \qquad (13)$$

We now compute $(C + D)/\pi(0, n_2)$ by using Relation (12):

$$\frac{C+D}{\pi(0,n_2)} = \frac{\Lambda_1}{\lambda_1 + \mu_1^0}\mu_1^0 + \frac{\lambda_1}{\Lambda_1 + \gamma_1^0} = \frac{\Lambda_1(\lambda_1 + 2\mu_1) + \lambda_1(\Lambda_1 + 2\gamma_1)}{2(\Lambda_1 + \gamma_1)}$$

$$= \frac{\lambda_1\Lambda_1 + \mu_1\Lambda_1 + \lambda_1\gamma_1}{\Lambda_1 + \gamma_1} = \lambda_1 + \frac{\mu_1\Lambda_1}{\lambda_e + \gamma_1} \quad (14)$$

Let us derive $E/\pi(0, n_2)$:

$$\frac{E}{\pi(0,n_2)} = \frac{\frac{\lambda_1\Lambda_1}{\Lambda_1+\gamma_1} + \mu_0\delta_{n_2=-1} + \mu\delta_{n_2<-1}}{\Lambda_2}\Lambda_2\delta_{n_2<0}$$

$$+ \frac{\Lambda_1\lambda_1}{\Lambda_1 + \gamma_1}\frac{1}{\Lambda_2 + \gamma_2^0\delta_{n_2=0} + \gamma_2\delta_{n_2\geq 1}}(\Lambda_2 + \gamma_2^0\delta_{n_2=0} + \gamma_2\delta_{n_2\geq 1})\delta_{n_2>0}$$

$$= \frac{\lambda_1\Lambda_1}{\Lambda_1 + \gamma_1} + \mu_2^0\delta_{n_2=-1} + \mu_2\delta_{n_<-1} \quad (15)$$

Notice that the sum $(C + D + E)/\pi(0, n_2) = \lambda_1 + \Lambda_1 + \mu_2^0\delta_{n_2=-1} + \mu_2\delta_{n_<-1}$. We now compute $(A + B + F)/\pi(0, n_2)$ to obtain the remaining terms of the LHS of Equation (11) divided by $\pi(0, n_2)$. By using Equation (13):

$$\frac{F}{\pi(0,n_2)} + \frac{A+B}{\pi(0,n_2)} = \frac{\Lambda_2}{\frac{\Lambda_1\lambda_1}{\Lambda_1+\gamma_1} + \mu_2^0\delta_{n_2=0} + \mu_2\delta_{n_2\leq -1}}(\mu_2\delta_{n_2\leq -1} + \mu_2^0\delta_{n_2=0})$$

$$+ \frac{\lambda_1\Lambda_1\Lambda_2}{\Lambda_1\lambda_1 + (\mu_2^0\delta_{n_2=0} + \mu_2\delta_{n_2\leq -1})(\Lambda_1 + \gamma_1)}\delta_{n_2\leq 0} + \Lambda_2\delta_{n_2>0} + \gamma_2^0\delta_{n_2=1} + \gamma_2\delta_{n_2>1}$$

$$= \Lambda_2 + \gamma_2^0\delta_{n_2=1} + \gamma_2\delta_{n_2>1}.$$

The last case is when $n_1 < 0$. In this case the GBE associated with states (n_1, n_2) have the form:

$$\pi(n_1, n_2)\big(\lambda_1 + \Lambda_1 + \mu_1^0\delta_{n_1=-1} + \mu_1\delta_{n_1<-1} + \Lambda_2 + \mu_2\delta_{n_2<-1} + \mu_2^0\delta_{n_2=-1}$$

$$+ \gamma_2^0\delta_{n_2=1} + \gamma_2\delta_{n_2>1}\big)$$

$$= \underbrace{\pi(n_1 - 1, n_2 - 1)\lambda_1}_{A} + \underbrace{\pi(n_1 - 1, n_2)\mu_1}_{B} + \underbrace{\pi(n_1 + 1, n_2)\Lambda_1}_{C}$$

$$+ \underbrace{\pi(n_1, n_2 - 1)(\mu_2^0\delta_{n_2=0} + \mu_2\delta_{n_2\leq -1})}_{D} + \underbrace{\pi(n_1, n_2 + 1)(\Lambda_2 + \gamma_2^0\delta_{n_2=0} + \gamma_2\delta_{n_2\geq 1})}_{E}$$

$$(16)$$

We divide the equation by $\pi(n_1, n_2)$ and consider the RHS:

$$\frac{A}{\pi(n_1,n_2)} = \frac{\Lambda_1}{\lambda_1 + \mu_1}\left(\frac{\Lambda_2\delta_{n_2\leq 0}}{\frac{\Lambda_1\lambda_1}{\Lambda_1+\gamma_1} + \mu_2^0\delta_{n_2=0} + \mu_2\delta_{n_2<0}}\right.$$

$$\left. + \frac{\Lambda_2 + \gamma_2^0\delta_{n_2=1} + \gamma_2\delta_{n_2>1}}{\frac{\Lambda_1\lambda}{\Lambda_1+\gamma_1}}\delta_{n_2\geq 1}\right)\lambda_1$$

$$= \frac{\Lambda_1}{\lambda_1 + \mu_1} \frac{\Lambda_2(\Lambda_1 + \gamma_1)\lambda_1\delta_{n_2 \le 0}}{\Lambda_1\lambda_1 + (\mu_2^0\delta_{n_2=0} + \mu_2\delta_{n_2<0})(\Lambda_1 + \gamma_1)}$$

$$+ \frac{\Lambda_1}{\lambda_1 + \mu_1} \frac{(\Lambda_1 + \gamma_1)(\Lambda_2 + \gamma_2^0\delta_{n_2=1} + \gamma_2\delta_{n_2>1})\lambda_1\delta_{n_2\ge1}}{\Lambda_1\lambda_1}$$

Recalling that by Assumptions 1 we have $\lambda_1 + \mu_1 = \Lambda_1 + \gamma_1$ this simplifies to:

$$\frac{A}{\pi(n_1,n_2)} = \frac{\Lambda_2\lambda_1\Lambda_1\delta_{n_2\le0}}{\Lambda_1\lambda_1 + (\mu_2^0\delta_{n_2=0} + \mu_2\delta_{n_2<0})(\Lambda_1 + \gamma_1)}$$

$$+ \Lambda_2\delta_{n_2\ge1} + \gamma_2^0\delta_{n_2=1} + \gamma_2\delta_{n_2>1} \quad (17)$$

$$\frac{B+E}{\pi(n_1,n_2)} = \frac{\Lambda_1\mu_1}{\lambda_1 + \mu_1}$$

$$+ \frac{\Lambda_1\lambda_1}{\Lambda_1 + \gamma_1} \frac{\Lambda_2 + \gamma_2^0\delta_{n_2=0} + \gamma_2\delta_{n_2\ge1}}{\Lambda_2 + \gamma_2^0\delta_{n_2=0} + \gamma_2\delta_{n_2\ge1}}\delta_{n_2\ge0}$$

$$+ \left(\frac{\Lambda_1\lambda_1}{\Lambda_1 + \gamma_1} + \mu_2^0\delta_{n_2=-1} + \mu_2\delta_{n_2<-1}\right)\frac{1}{\Lambda_2}\Lambda_2\delta_{n_2<0}$$

$$\frac{\Lambda_1\mu_1}{\lambda_1 + \mu_1} + \frac{\Lambda_1\lambda_1}{\Lambda_1 + \gamma_1} + \mu_2^0\delta_{n_2=-1} + \mu_2\delta_{n_2<-1}$$

$$= \Lambda_1 + \mu_2^0\delta_{n_2=-1} + \mu_2\delta_{n_2<-1} \quad (18)$$

$$\frac{C}{\pi(n_1,n_2)} = \frac{\lambda_1 + \mu_1\delta_{n_1<-1} + \mu_1^0\delta_{n_1=-1}}{\Lambda_1}\Lambda_1 = \lambda_1 + \mu_1\delta_{n_1<-1} + \mu_1^0\delta_{n_1=-1} \quad (19)$$

$$\frac{D}{\pi(n_1,n_2)} = \frac{\Lambda_2(\Lambda_1 + \gamma_1)(\mu_2^0\delta_{n_2=0} + \mu_2\delta_{n_2\le-1})\delta_{n_2\le0}}{\Lambda_1\lambda_1 + (\mu_2^0\delta_{n_2=0} + \mu_2\delta_{n_2<0})(\Lambda_1 + \gamma_1)} \quad (20)$$

The analysis of the global balance equation system is concluded by observing that summing Equations (17), (18), (19), (20), we obtain the LHS of the balance equation (16) as required.

As regards the derivation of the normalising constant it is sufficient to compute the sum of the geometric series given by summing Equation (2) over the state space $(-\infty, +\infty)$.   □

Measuring the Distance Between MAPs and Some Applications

Gábor Horváth[1,2]([envelope])

[1] Department of Networked Systems and Services,
Budapest University of Technology and Economics, Budapest, Hungary
[2] MTA-BME Information Systems Research Group, Magyar Tudósok krt. 2,
Budapest 1117, Hungary
ghorvath@hit.bme.hu

Abstract. This paper provides closed form expressions for the squared distance between the joint density functions of k successive inter-arrival times of two MAPs. The squared distance between the autocorrelation functions of two MAPs is expressed in a closed form as well.

Based on these results a simple procedure is developed to approximate a RAP by a MAP, in order to reduce the number of phases or to obtain a Markovian representation.

1 Introduction

MAPs (Markovian Arrival Processes) and their generalizations, RAPs (Rational Arrival Processes) are versatile modeling tools in various fields of performance evaluation. They represent a dense class of point processes ([1]), and at the same time they are easy to work with: several important statistical properties can be expressed in a simple closed form, they exhibit many closeness properties, queues involving MAP arrival and/or service process can be solved efficiently, etc.

In the last decades considerable research effort has been spent to approximate various point processes by MAPs to take the advantage of their technical simplicity. Matching and fitting methods have been developed to construct MAPs based on empirical measurement traces, or based on point processes like departure processes of queues, etc. However, the MAPs or RAPs produced by some of these procedures might not be ready for use immediately. There are situations when compactness (in terms of the number of states) and the Markovian representation is important.

In order to develop procedures to compress a MAP and/or to obtain a Markovian approximation of a RAP, it is necessary to define distance functions which measure how "close" two RAPs are to each other. Since this distance function is evaluated repetitively in an optimization procedure, it must be efficient to evaluate.

In this paper we show that the squared distance between the joint density functions of k successive inter-arrival times of two MAPs can be expressed in a closed form. Furthermore, the squared distance between the autocorrelation

© Springer International Publishing Switzerland 2015
M. Gribaudo et al. (Eds.): ASMTA 2015, LNCS 9081, pp. 100–114, 2015.
DOI: 10.1007/978-3-319-18579-8_8

functions can be expressed in a closed form as well. Based on these results a simple procedure is developed to approximate a RAP by a MAP, and some possible applications are also provided.

The rest of the paper is organized as follows. Section 2 introduces the notations and the main properties of MAPs and RAPs used in the paper. Section 3 presents how the distance between two MAPs is calculated. The RAP approximation procedure is developed in Section 4. Finally, Section 5 demonstrates how the results are applied for the approximation of the departure process of a MAP/MAP/1 queue.

2   Markovian Arrival Processes

A Markovian Arrival Process (MAP, [7]) with N phases is given by two $N \times N$ matrices, $\boldsymbol{D_0}$ and $\boldsymbol{D_1}$. The sum $\boldsymbol{D} = \boldsymbol{D_0} + \boldsymbol{D_1}$ is the generator of an irreducible continuous time Markov chain (CTMC) with N states, which is the background process of the MAP. Matrix $\boldsymbol{D_1}$ contains the rates of those phase transitions which are accompanied by an arrival, and the off-diagonal entries of $\boldsymbol{D_0}$ are the rates of internal phase transitions.

The phase process embedded at arrival instants plays an important role in the analysis of MAPs. This phase process is a discrete time Markov chain whose transition probability matrix is $\boldsymbol{P} = (-\boldsymbol{D_0})^{-1}\boldsymbol{D_1}$. The stationary probability vector of the embedded process is denoted by α, it is the unique solution to linear equations $\alpha\boldsymbol{P} = \alpha, \alpha\mathbb{1} = 1$.

The joint density function of k consecutive inter-arrival times $\mathcal{X}_1, \mathcal{X}_2, \ldots \mathcal{X}_k$ is given by

$$f_k(x_1, x_2, \ldots, x_k) = \alpha e^{\boldsymbol{D_0}x_1}\boldsymbol{D_1} \cdot e^{\boldsymbol{D_0}x_2}\boldsymbol{D_1} \cdots e^{\boldsymbol{D_0}x_k}\boldsymbol{D_1}\mathbb{1}. \tag{1}$$

The lag-k autocorrelation of the inter-arrival times is matrix-geometric, and can be expressed as

$$\begin{aligned}
\rho_k &= \frac{E(\mathcal{X}_1\mathcal{X}_{k+1}) - E(\mathcal{X}_1)^2}{E(\mathcal{X}_1^2) - E(\mathcal{X}_1)^2} \\
&= \frac{\alpha(-\boldsymbol{D_0})^{-1}\boldsymbol{P}^k(-\boldsymbol{D_0})^{-1}\mathbb{1} - \alpha(-\boldsymbol{D_0})^{-1}\mathbb{1} \cdot \alpha(-\boldsymbol{D_0})^{-1}\mathbb{1}}{\sigma^2} \\
&= \frac{1}{\sigma^2}\alpha(-\boldsymbol{D_0})^{-1}(\boldsymbol{P} - \mathbb{1}\alpha)^k(-\boldsymbol{D_0})^{-1}\mathbb{1}
\end{aligned} \tag{2}$$

for $k > 0$, and it is $\rho_0 = 1$ for $k = 0$. σ^2 denotes the variance of the inter-arrival times. In (2) we exploited that $\boldsymbol{P}^k - \mathbb{1}\alpha = (\boldsymbol{P} - \mathbb{1}\alpha)^k$ holds for $k > 0$ (notice however that it does not hold for $k = 0$).

Rational Arrival Processes (RAPs) are generalizations of MAPs, which do not have the Markovian restrictions. The $\boldsymbol{D_0}, \boldsymbol{D_1}$ matrices of RAPs can have arbitrary entries, the only restriction is that the joint density function must be valid. However, without loss of generality we assume that $(\boldsymbol{D_0} + \boldsymbol{D_1})\mathbb{1} = \mathbb{1}$ holds throughout the paper. By the appropriate similarity transformation all RAPs can be transformed to this form ([8]), and several authors apply this assumption to make the corresponding derivations simpler.

Getting rid of the Markovian restrictions makes RAPs easier to use than MAPs in several situations, but checking that a RAP is a valid stochastic process is hard (apart from the case when the transformation to a Markovian representation is successful).

Since this paper is on measuring the distance between two MAPs/RAPs, we are going to leave the traditional (D_0, D_1) notation of the MAP matrices behind and use different letters instead.

3   Efficient Calculation of the Distance Between Two MAPs

3.1   The Distance Between the Joint Density Functions of Two MAPs

Let us consider two MAPs, $\boldsymbol{A} = (\boldsymbol{A_0}, \boldsymbol{A_1})$ and $\boldsymbol{B} = (\boldsymbol{B_0}, \boldsymbol{B_1})$. The squared difference of the joint density of the inter-arrival times up to lag-k is defined by

$$
\mathcal{D}_k\{\boldsymbol{A}, \boldsymbol{B}\} = \int_0^\infty \cdots \int_0^\infty \int_0^\infty \left(\alpha_{\mathcal{A}} e^{\boldsymbol{A_0} x_1} \boldsymbol{A_1} \cdots e^{\boldsymbol{A_0} x_{k-1}} \boldsymbol{A_1} \cdot e^{\boldsymbol{A_0} x_k} \boldsymbol{A_1} \mathbb{1} \right.
$$
$$
\left. - \alpha_{\mathcal{B}} e^{\boldsymbol{B_0} x_1} \boldsymbol{B_1} \cdots e^{\boldsymbol{B_0} x_{k-1}} \boldsymbol{B_1} \cdot e^{\boldsymbol{B_0} x_k} \boldsymbol{B_1} \mathbb{1} \right)^2 dx_1 \ldots dx_{k-1}\, dx_k,
\tag{3}
$$

where $\alpha_{\mathcal{A}}$ and $\alpha_{\mathcal{B}}$ denote the stationary phase distribution of MAPs $\boldsymbol{A}$ and $\boldsymbol{B}$ at arrival instants. The square term expands to

$$
\mathcal{D}_k\{\boldsymbol{A}, \boldsymbol{B}\} = L_k(\boldsymbol{A}, \boldsymbol{A}) - 2L_k(\boldsymbol{A}, \boldsymbol{B}) + L_k(\boldsymbol{B}, \boldsymbol{B}),
\tag{4}
$$

where $L_k(\boldsymbol{A}, \boldsymbol{B})$ represents the integral

$$
L_k(\boldsymbol{A}, \boldsymbol{B}) = \int_0^\infty \cdots \int_0^\infty \int_0^\infty \alpha_{\mathcal{A}} e^{\boldsymbol{A_0} x_1} \boldsymbol{A_1} \cdots e^{\boldsymbol{A_0} x_{k-1}} \boldsymbol{A_1} \cdot e^{\boldsymbol{A_0} x_k} \boldsymbol{A_1} \mathbb{1}
$$
$$
\cdot \alpha_{\mathcal{B}} e^{\boldsymbol{B_0} x_1} \boldsymbol{B_1} \cdots e^{\boldsymbol{B_0} x_{k-1}} \boldsymbol{B_1} \cdot e^{\boldsymbol{B_0} x_k} \boldsymbol{B_1} \mathbb{1}\, dx_1 \ldots dx_{k-1}\, dx_k.
\tag{5}
$$

This integral can be evaluated in an efficient way, by successive solution of (Sylvester-type) linear equations, as stated by the following theorem.

Theorem 1. $L_k(\boldsymbol{A}, \boldsymbol{B})$ *can be expressed by*

$$
L_k(\boldsymbol{A}, \boldsymbol{B}) = \mathbb{1}^T \boldsymbol{B_1}^T \cdot \boldsymbol{Y_k} \cdot \boldsymbol{A_1} \mathbb{1},
\tag{6}
$$

where matrix $\boldsymbol{Y_k}$ is the solution of the recursive Sylvester equation

$$
\begin{cases}
-\boldsymbol{B_1}^T \boldsymbol{Y_{k-1}} \boldsymbol{A_1} = \boldsymbol{B_0}^T \boldsymbol{Y_k} + \boldsymbol{Y_k} \boldsymbol{A_0} & \text{for } k > 1, \\
-\alpha_{\mathcal{B}}^T \alpha_{\mathcal{A}} = \boldsymbol{B_0}^T \boldsymbol{Y_1} + \boldsymbol{Y_1} \boldsymbol{A_0} & \text{for } k = 1.
\end{cases}
\tag{7}
$$

Proof. We start by transforming (5) as

$$
L_k(\mathcal{A}, \mathcal{B}) = \int_0^\infty \cdots \int_0^\infty \int_0^\infty \mathbb{1}^T \boldsymbol{B}_1^T e^{\boldsymbol{B}_0^T x_k} \boldsymbol{B}_1^T e^{\boldsymbol{B}_0^T x_{k-1}} \cdots \boldsymbol{B}_1^T e^{\boldsymbol{B}_0^T x_1} \alpha_\mathcal{B}^T
$$
$$
\cdot \, \alpha_\mathcal{A} e^{\boldsymbol{A}_0 x_1} \boldsymbol{A}_1 \cdots e^{\boldsymbol{A}_0 x_{k-1}} \boldsymbol{A}_1 \cdot e^{\boldsymbol{A}_0 x_k} \boldsymbol{A}_1 \mathbb{1} \, dx_1 \ldots dx_{k-1} \, dx_k
$$
$$
= \mathbb{1}^T \boldsymbol{B}_1^T \left(\int_0^\infty \cdots \int_0^\infty \int_0^\infty e^{\boldsymbol{B}_0^T x_k} \boldsymbol{B}_1^T e^{\boldsymbol{B}_0^T x_{k-1}} \cdots \boldsymbol{B}_1^T e^{\boldsymbol{B}_0^T x_1} \alpha_\mathcal{B}^T \right. \tag{8}
$$
$$
\left. \cdot \, \alpha_\mathcal{A} e^{\boldsymbol{A}_0 x_1} \boldsymbol{A}_1 \cdots e^{\boldsymbol{A}_0 x_{k-1}} \boldsymbol{A}_1 \cdot e^{\boldsymbol{A}_0 x_k} \, dx_1 \ldots dx_{k-1} \, dx_k \right) \cdot \boldsymbol{A}_1 \mathbb{1}.
$$

Let us denote the term in the parenthesis by $\boldsymbol{Y}_k$. For $k > 1$, separating the first and the last terms leads to the recursion

$$
\boldsymbol{Y}_k = \int_0^\infty e^{\boldsymbol{B}_0^T x_k} \cdot \boldsymbol{B}_1^T \left(\int_0^\infty \cdots \int_0^\infty e^{\boldsymbol{B}_0^T x_{k-1}} \boldsymbol{B}_1^T \cdots \boldsymbol{B}_1^T e^{\boldsymbol{B}_0^T x_1} \alpha_\mathcal{B}^T \right.
$$
$$
\left. \cdot \, \alpha_\mathcal{A} e^{\boldsymbol{A}_0 x_1} \boldsymbol{A}_1 \cdots e^{\boldsymbol{A}_0 x_{k-1}} \boldsymbol{A}_1 \, dx_1 \ldots dx_{k-1} \right) \boldsymbol{A}_1 \cdot e^{\boldsymbol{A}_0 x_k} \, dx_k \tag{9}
$$
$$
= \int_0^\infty e^{\boldsymbol{B}_0^T x_k} \boldsymbol{B}_1^T \cdot \boldsymbol{Y}_{k-1} \cdot \boldsymbol{A}_1 e^{\boldsymbol{A}_0 x_k} \, dx_k,
$$

which is the solution of Sylvester equation $-\boldsymbol{B}_1^T \boldsymbol{Y}_{k-1} \boldsymbol{A}_1 = \boldsymbol{B}_0^T \boldsymbol{Y}_k + \boldsymbol{Y}_k \boldsymbol{A}_0$. The equation for $k = 1$ is obtained similarly. □

Note that the solution of (7) is always unique as matrices $\boldsymbol{A}_0$ and $\boldsymbol{B}_0$ are sub-generators.

3.2 The Distance Between the Lag Autocorrelation Functions

The squared distance between the lag autocorrelation functions of MAP $\mathcal{A}$ and $\mathcal{B}$ is computed by

$$
\mathcal{D}_{\mathrm{acf}}\{\mathcal{A}, \mathcal{B}\} = \sum_{i=0}^\infty (\rho_i^{(A)} - \rho_i^{(B)})^2
$$
$$
= \sum_{i=1}^\infty \left(\frac{1}{\sigma_\mathcal{A}^2} \alpha_\mathcal{A} (-\boldsymbol{A}_0)^{-1} (\boldsymbol{P}_A - \mathbb{1}\alpha_\mathcal{A})^i (-\boldsymbol{A}_0)^{-1} \mathbb{1} \right. \tag{10}
$$
$$
\left. - \frac{1}{\sigma_\mathcal{B}^2} \alpha_\mathcal{B} (-\boldsymbol{B}_0)^{-1} (\boldsymbol{P}_B - \mathbb{1}\alpha_\mathcal{B})^i (-\boldsymbol{B}_0)^{-1} \mathbb{1} \right)^2,
$$

where $\sigma_\mathcal{A}^2$ ($\sigma_\mathcal{B}^2$) denotes the variance of the inter-arrival times of MAP $\mathcal{A}$ ($\mathcal{B}$), respectively. Expanding the square term leads to

$$
\mathcal{D}_{\mathrm{acf}}\{\mathcal{A}, \mathcal{B}\} = \frac{1}{\sigma_\mathcal{A}^4} \left(M(\mathcal{A}, \mathcal{A}) - m_2^{(A)^2}/4 \right)
$$
$$
- 2 \frac{1}{\sigma_\mathcal{A}^2 \sigma_\mathcal{B}^2} \left(M(\mathcal{A}, \mathcal{B}) - m_2^{(A)} m_2^{(B)}/4 \right) \tag{11}
$$
$$
+ \frac{1}{\sigma_\mathcal{B}^4} \left(M(\mathcal{B}, \mathcal{B}) - m_2^{(B)^2}/4 \right),
$$

where $m_2^{(A)}$ and $m_2^{(B)}$ denote the second moment of the inter-arrival times of MAP $\mathcal{A}$ and $\mathcal{B}$, while matrix $M(\mathcal{A}, \mathcal{B})$ represents the sum

$$M(\mathcal{A}, \mathcal{B}) = \sum_{i=0}^{\infty} \alpha_{\mathcal{A}} (-A_0)^{-1} (P_{\mathcal{A}} - \mathbb{1}\alpha_{\mathcal{A}})^i (-A_0)^{-1} \mathbb{1} \cdot \\ \cdot \alpha_{\mathcal{B}} (-B_0)^{-1} (P_{\mathcal{B}} - \mathbb{1}\alpha_{\mathcal{B}})^i (-B_0)^{-1} \mathbb{1}. \tag{12}$$

The terms involving the second moments in (11) are necessary since the sum goes from $i = 1$ in (10) and it goes from $i = 0$ in (12). Term 0 of $M(\mathcal{A}, \mathcal{B})$ equals $m_2^{(A)}/2 \cdot m_2^{(B)}/2$.

The next theorem provides the solution of matrix $M(\mathcal{A}, \mathcal{B})$.

Theorem 2. *Matrix $M(\mathcal{A}, \mathcal{B})$ is obtained by*

$$M(\mathcal{A}, \mathcal{B}) = \alpha_{\mathcal{A}} (-A_0)^{-1} \cdot X \cdot (-B_0)^{-1} \mathbb{1}, \tag{13}$$

where X is the unique solution to the discrete Sylvester equation

$$(P_{\mathcal{A}} - \mathbb{1}\alpha_{\mathcal{A}}) \cdot X \cdot (P_{\mathcal{B}} - \mathbb{1}\alpha_{\mathcal{B}}) - X + (-A_0)^{-1} \mathbb{1}\alpha_{\mathcal{B}} (-B_0)^{-1} = 0. \tag{14}$$

Proof. Matrices $P_{\mathcal{A}} - \mathbb{1}\alpha_{\mathcal{A}}$ and $P_{\mathcal{B}} - \mathbb{1}\alpha_{\mathcal{B}}$ are stable, since the subtraction of $\mathbb{1}\alpha_{\mathcal{A}}$ and $\mathbb{1}\alpha_{\mathcal{B}}$ removes the eigenvalue of 1 which matrices $P_{\mathcal{A}}$ and $P_{\mathcal{B}}$ originally had. Hence we can utilize that the solution of the sum $X = \sum_{i=0}^{\infty} A^i C B^i$ satisfies the discrete Sylvester equation $AXB - X + C = 0$. □

4   Application: Approximating a RAP with a MAP

Having results for measuring the distance between two RAPs or MAPs can be useful in many situations by themselves. In this section we use them as distance functions in an optimization problem. We develop a simple procedure to obtain a MAP that approximates the behavior of a given RAP. Two possible applications of this procedure are as follows.

- Several matching procedures produce a RAP which does not have a Markovian representation, or which is not even a valid stochastic process (the joint density is negative at some points). The presented procedure returns a valid MAP that is as close as possible to the target RAP.
- Several performance models involve huge MAPs which make the analysis too slow and numerically challenging. With the presented procedure it is possible to compress these large MAPs by constructing small replacements that are easier to work with.

Throughout this section the target RAP is denoted by $\mathcal{A} = (A_0, A_1)$ and the approximating one by $\mathcal{B} = (B_0, B_1)$.

4.1   Obtaining Matrix B_1 Given that $\alpha_\mathcal{B}$ and B_0 Are Known

Given that $\alpha_\mathcal{B}$ and B_0 are already available (see later in Section 4.2) matrix B_1
it obtained

- either to minimize $\mathcal{D}_k\{A, B\}$ up to a given k,
- or to minimize $\mathcal{D}_{\mathrm{acf}}\{A, B\}$.

According to the following theorem, optimizing the squared distance of the
lag-1 joint density function $\mathcal{D}_2\{\mathcal{A}, \mathcal{B}\}$ is especially efficient.

Theorem 3. *Given that $\alpha_\mathcal{B}$ and B_0 are available, matrix B_1 minimizing*
$\mathcal{D}_2\{\mathcal{A}, \mathcal{B}\}$ *is the solution of the quadratic program*

$$\min_{B_1} \left\{ vec\langle B_1 \rangle^T (W_{BB} \otimes Y_{BB}) vec\langle B_1 \rangle - 2vec\langle A_1 \rangle^T (W_{AB} \otimes Y_{AB}) vec\langle B_1 \rangle \right\}$$
(15)

subject to

$$\left(I \otimes \alpha_\mathcal{B}(-B_0)^{-1} \right) vec\langle B_1 \rangle = \alpha_\mathcal{A},$$
(16)

$$(\mathbb{1}^T \otimes I) vec\langle B_1 \rangle = -B_0 \mathbb{1}.$$
(17)

Matrices W_{AB}, W_{BB}, Y_{AB} *and* Y_{BB} *are the solutions to Sylvester equations*

$$A_0 W_{AB} + W_{AB} B_0^T = -A_0 \mathbb{1} \cdot \mathbb{1}^T B_0^T,$$
(18)

$$B_0 W_{BB} + W_{BB} B_0^T = -B_0 \mathbb{1} \cdot \mathbb{1}^T B_0^T,$$
(19)

$$A_0{}^T Y_{AB} + Y_{AB} B_0 = -\alpha_\mathcal{A}^T \cdot \alpha_\mathcal{B},$$
(20)

$$B_0^T Y_{BB} + Y_{BB} B_0 = -\alpha_\mathcal{B}^T \cdot \alpha_\mathcal{B}.$$
(21)

Proof. Let us first apply the $vec\langle\rangle$ (column stacking) operator on (6) at $k = 2$.
Utilizing the identity $vec\langle AXB \rangle = (B^T \otimes A)vec\langle X \rangle$ for compatible matrices
A, B, X and the identity $vec\langle u^T v \rangle = (v^T \otimes u^T)$ for row vectors u and v (see [9]).
We get

$$vec\langle L_2(\mathcal{A}, \mathcal{B}) \rangle = (\mathbb{1}^T A_0^T \otimes \mathbb{1}^T B_0^T) \cdot vec\langle Y_2 \rangle = vec\langle B_0 \mathbb{1} \cdot \mathbb{1}^T A_0^T \rangle^T \cdot vec\langle Y_2 \rangle. \quad (22)$$

Applying the $vec\langle\rangle$ operator on both sides of (7) and using $vec\langle AXB \rangle = (B^T \otimes A)vec\langle X \rangle$ again leads to

$$-(I \otimes B_1^T Y_1)vec\langle A_1 \rangle = (I \otimes B_0^T)vec\langle Y_2 \rangle + (A_0{}^T \otimes I)vec\langle Y_2 \rangle, \quad (23)$$

from which $vec\langle Y_2 \rangle$ is expressed by

$$vec\langle Y_2 \rangle = (-A_0{}^T \oplus B_0^T)^{-1}(I \otimes B_1^T)(I \otimes Y_{AB})vec\langle A_1 \rangle, \quad (24)$$

since $Y_1 = Y_{AB}$. Thus we have

$$vec\langle L_2(\mathcal{A}, \mathcal{B}) \rangle = \underbrace{vec\langle B_0 \mathbb{1} \cdot \mathbb{1}^T A_0^T \rangle^T (-A_0{}^T \oplus B_0^T)^{-1}}_{vec\langle W_{AB} \rangle^T} (I \otimes B_1^T)(I \otimes Y_{AB})vec\langle A_1 \rangle,$$

(25)

where we recognized that the transpose of $\text{vec}\langle W_{AB}\rangle$ expressed from (18) matches the first two terms of the expression. Using the identities of the $\text{vec}\langle\rangle$ operator yields

$$\text{vec}\langle W_{AB}\rangle^T (I \otimes B_1^T) = \text{vec}\langle B_1^T W_{AB}\rangle^T = \text{vec}\langle B_1\rangle^T (W_{AB} \otimes I). \qquad (26)$$

Finally, putting together (25) and (26) gives

$$\text{vec}\langle L_2(\mathcal{A}, \mathcal{B})\rangle = \text{vec}\langle B_1\rangle^T (W_{AB} \otimes Y_{AB})\text{vec}\langle A_1\rangle. \qquad (27)$$

From the components of $\mathcal{D}_2\{\mathcal{A}, \mathcal{B}\}$ (see (4)) $L_2(\mathcal{A}, \mathcal{A})$ plays no role in the optimization as it does not depend on B_1, the term $L_2(\mathcal{A}, \mathcal{B})$ yields the linear term in (15) according to (27), and $L_2(\mathcal{B}, \mathcal{B})$ introduces the quadratic term, based on (27) after replacing $\mathcal{A}$ by $\mathcal{B}$.

According to the first constraint (16) and the second constraint (17) the solution must satisfy $\alpha_{\mathcal{B}}(-B_0)^{-1}B_1 = \alpha_{\mathcal{B}}$ and $B_1 \mathbb{1} = -B_0 \mathbb{1}$, respectively. $\square$

Theorem 4. *Matrix $W_{BB} \otimes Y_{BB}$ is positive definite, thus the quadratic optimization problem of Theorem 3 is convex.*

Proof. If W_{BB} and Y_{BB} are positive definite, then their Kronecker product is positive definite as well. First we show that matrix Y_{BB} is positive definite, thus $z Y_{BB} z^T > 0$ holds for any non-zero row vector z. Since Y_{BB} is the solution of a Sylvester equation, we have that $Y_{BB} = \int_0^\infty e^{B_0^T x} \alpha_{\mathcal{B}}^T \cdot \alpha_{\mathcal{B}} e^{B_0 x}\, dx$. Hence

$$z Y_{BB} z^T = \int_0^\infty z e^{B_0^T x} \alpha_{\mathcal{B}}^T \cdot \alpha_{\mathcal{B}} e^{B_0 x} z^T\, dx = \int_0^\infty \left(\alpha_{\mathcal{B}} e^{B_0 x} z^T\right)^2 dx, \qquad (28)$$

which can not be negative, furthermore, apart from a finite number of x values $\alpha_{\mathcal{B}} e^{B_0 x} z^T$ can not be zero either. Thus, the integral is always strictly positive.

The positive definiteness of matrix W_{BB} can be proven similarly. $\square$

Being able to formalize the optimization of $\mathcal{D}_2\{\mathcal{A}, \mathcal{B}\}$ as a quadratic programming problem means that obtaining the optimal matrix B_1 is efficient: it is fast, and there is a single optimum which is always found.

If we intend to take higher lag joint density differences also into account, the objective function is $\mathcal{D}_k\{\mathcal{A}, \mathcal{B}\}$, which is not quadratic for $k > 2$. However, our numerical experience is that the built-in non-linear optimization tool in MATLAB, called `fmincon` is able to return the solution matrix B_1 quickly, independent of the initial point of the optimization. We have a strong suspicion that the returned solution is the global optimum, however we can not prove the convexity of the objective function formally.

It is also possible to use $\mathcal{D}_{acf}\{\mathcal{A}, \mathcal{B}\}$ as the objective function of the optimization problem, when looking for matrix B_1 that minimizes the squared difference of the autocorrelation function. We found that `fmincon` is rather prone to the initial point in this case. Repeated running with different random initial points was required to obtain the best solution.

4.2   Approximating a RAP

The proposed procedure consists of two steps:

1. obtaining the phase-type (PH) representation of the inter-arrival times, that provides vector α_B and matrix B_0;
2. obtaining the optimal B_1 matrix such that the correlation structure of the target RAP is captured as accurately as possible.

Section 4.1 describes how step 2 is performed.

For step 1, any phase-type fitting method can be applied. To solve this problem [3] develops a moment matching method that returns a hyper-exponential distribution of order N based on $2N - 1$ moments, if it is possible. An other solution published in [6] is based on a hyper-Erlang distribution, which always succeeds if an appropriately large Erlang order is chosen.

Our method of choice, however, is a slight modification of [5], which is the generalization of the former two. It constructs PH distributions from feedback Erlang blocks (FEBs), where each FEB implements an eigenvalue of the target distribution. With FEBs it is possible to represent complex eigenvalues as well, as opposed to the previously mentioned methods that operate on hyper-exponential and hyper-Erlang distributions. The original method in [5] puts the FEBs in a row, which is not appropriate for our goals, since there is only a single absorbing state, implying that matrix B_1 can have only a single non-zero row, thus no correlation can be realized. However, the original method can be modified in a straight forward way to return a hyper-FEB structure. A key step of [5] is the solution of a polynomial system of equations, which can have several solutions, providing several valid α_B, B_0 pairs. Our RAP approximation procedure performs the optimization of matrix B_1 with all of these solutions, and picks the best one among them.

4.3   Numerical Examples

In the first numerical example we extract 7 marginal moments and 9 lag-1 joint moments from a measurement trace containing inter-arrival times of real data traffic[1], and create a RAP of order 4 with the method published in [10]. The obtained matrices are as follows:

$$
A_0 = \begin{bmatrix} -0.579 & -0.402 & -0.364 & -0.348 \\ -0.368 & -0.205 & -0.315 & -0.36 \\ 1.32 & -0.845 & 0.701 & 1.13 \\ -1.7 & 0.3 & -1.14 & -1.52 \end{bmatrix}, \quad A_1 = \begin{bmatrix} 0.576 & 0.262 & 0.41 & 0.446 \\ 0.168 & 0.501 & 0.313 & 0.266 \\ 0.29 & -1.69 & -0.598 & -0.302 \\ 0.292 & 1.94 & 1.03 & 0.786 \end{bmatrix}.
$$

The RAP characterized by $\mathcal{A} = (A_0, A_1)$ is, however, not a valid stochastic process as the joint density given by (1) is negative since $f_2(0.5, 8) = -0.000357$. This RAP is the target of our approximation in this section.

---

[1] We used the BC-pAug89 trace, http://ita.ee.lbl.gov/html/contrib/BC.html. While this is a fairly old trace, it is often used for testing PH and MAP fitting methods, it became like a benchmark.

Let us now construct a MAP $\mathcal{B}^{(1)} = (\mathbf{B}_0^{(1)}, \mathbf{B}_1^{(1)})$ which minimizes the squared distance of the lag-1 joint density with $\mathcal{A}$. The distribution of the inter-arrival times, characterized by $\alpha_B, \mathbf{B}_0^{(1)}$ are obtained by the modified moment matching method of [5], and matrix $\mathbf{B}_1^{(1)}$ has been determined by the quadratic program provided by Theorem 3. The matrices of the MAP are

$$\mathbf{B}_0^{(1)} = \begin{bmatrix} -0.074 & 0 & 0 & 0 & 0 \\ 0 & -0.27 & 0.27 & 0 & 0 \\ 0 & 0 & -0.27 & 0.27 & 0 \\ 0 & 0 & 0 & -0.27 & 0 \\ 0 & 0 & 0 & 0 & -1.2 \end{bmatrix}, \mathbf{B}_1^{(1)} = \begin{bmatrix} 0.0065 & 0.024 & 0 & 5.5 \cdot 10^{-8} & 0.044 \\ 0 & 0 & 0 & 0 & 0 \\ 0 & 0 & 0 & 0 & 0 \\ 0.017 & 0.086 & 0 & 0 & 0.17 \\ 0 & 0.012 & 0 & 0 & 1.2 \end{bmatrix},$$

and the squared distance in the lag-1 joint pdf is $\mathcal{D}_2\{\mathcal{A}, \mathcal{B}^{(1)}\} = 0.000105$. The quadratic program has been solved by MATLAB is less than a second. Next, we repeat the same procedure, but instead of focusing on the lag-1 distance, we optimize on the squared distance of the joint pdf up to lag-10. This can not be formalized as a quadratic program any more, but the optimization is still fast, lasting only 1-2 seconds. In this case the hyper-exponential distribution provided the best results $(\mathcal{D}_{11}\{\mathcal{A}, \mathcal{B}^{(10)}\} = 4.37 \cdot 10^{-5})$. The matrices are

$$\mathbf{B}_0^{(10)} = \begin{bmatrix} -0.0519 & 0 & 0 \\ 0 & -0.151 & 0 \\ 0 & 0 & -1.24 \end{bmatrix}, \quad \mathbf{B}_1^{(10)} = \begin{bmatrix} 10^{-6} & 0.0519 & 10^{-6} \\ 10^{-6} & 0.151 & 0.000465 \\ 0.000129 & 10^{-6} & 1.24 \end{bmatrix}.$$

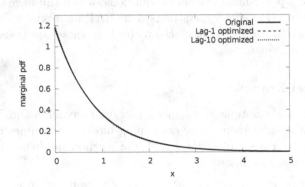

Fig. 1. Comparison of the density functions of the marginal distribution

To evaluate the quality of the approximation Figure 1 compares the marginal density functions of $\mathcal{A}, \mathcal{B}^{(1)}$ and $\mathcal{B}^{(10)}$. The plots are close to each other, the approximation is relatively accurate. To demonstrate that the lag-1 joint densities are also accurate, Figure 2 depicts them at $x_2 = 0.5, 1$ and 1.5.

In the next experiment the objective is the squared distance of the lag-k autocorrelation function. As before, the input RAP is $\mathcal{A}$, but now the approximation procedure has to minimize $\mathcal{D}_{\text{acf}}\{\mathcal{A}, \mathcal{B}^{(\rho)}\}$ which is given in a closed form

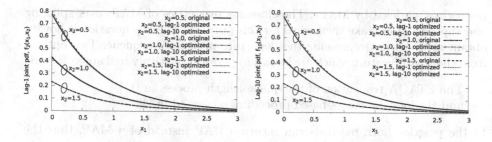

Fig. 2. Comparison of the lag-1 joint density functions

by (11) and Theorem 2. According to our experience the result of the optimization is rather prone to the initial point. The best result from 10 trials is given by matrices

$$
B_0^{(\rho)} = \begin{bmatrix} -0.0851 & 0.0851 & 0 & 0 & 0 & 0 \\ 0 & -0.0851 & 0 & 0 & 0 & 0 \\ 0 & 0 & -0.267 & 0.267 & 0 & 0 \\ 0 & 0 & 0 & -0.267 & 0.267 & 0 \\ 0 & 0 & 0 & 0 & -0.267 & 0 \\ 0 & 0 & 0 & 0 & 0 & -1.2 \end{bmatrix}, B_1^{(\rho)} = \begin{bmatrix} 0 & 0 & 0 & 0 & 0 & 0 \\ 0 & 0 & 0.0485 & 0 & 0 & 0.0366 \\ 0 & 0 & 0 & 0 & 0 & 0 \\ 0 & 0 & 0 & 0 & 0 & 0 \\ 0 & 0 & 0.0965 & 0 & 0 & 0.1705 \\ 0.0004 & 0 & 0.0117 & 0 & 0 & 1.1885 \end{bmatrix}.
$$

and the corresponding autocorrelation function is depicted in Figure 3. The squared distance between the autocorrelation functions is $\mathcal{D}_{\mathrm{acf}}\{\mathcal{A}, \mathcal{B}^{(\rho)}\} = 0.00237$.

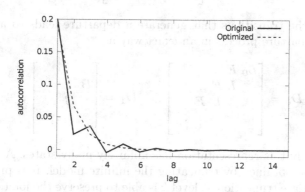

Fig. 3. Comparison of the autocorrelation functions

5 Application: Approximating the Departure Process of a MAP/MAP/1 Queue by a MAP

A popular approach for the analysis of the network of MAP/MAP/1 queues is the so called traffic based decomposition, where the internal traffic in the network is modeled by MAPs. The closeness properties of MAPs over splitting and superposition make them ideal for this purpose. The key question is how to obtain a MAP that represents the departure process of a queue. Two options from the past literature which are known to perform relatively well are as follows:

- The ETAQA truncation of the queue length process in [11],
- and the joint moments based procedure presented in [4].

In the practice both methods can return a RAP instead of a MAP, thus the procedure described in Section 4 becomes relevant.

5.1   Introduction to the Departure Process Analysis

The MAP/MAP/1 queue is a subclass of QBD queues, which are characterized by four matrices, B, F, L and L_0. Matrices B and F consist of phase transition rates accompanied by service and arrival events, respectively, while matrices L_0 and L correspond to the internal transitions when the queue is at level 0 and at level above zero. The generator matrix of the CTMC keeping track of the number of jobs in the queue and the phase of the system has a tri-diagonal structure given by

$$Q = \begin{bmatrix} L_0 & F & & \\ B & L & F & \\ & B & L & F \\ & & \ddots & \ddots & \ddots \end{bmatrix}. \tag{29}$$

Separating the transitions that generate a departure leads to a MAP that captures the departure process in an exact way as

$$D_0 = \begin{bmatrix} L_0 & F & & \\ & L & F & \\ & & L & F \\ & & & \ddots & \ddots \end{bmatrix}, \quad D_1 = \begin{bmatrix} B & & \\ & B & \\ & & \ddots \end{bmatrix}, \tag{30}$$

but unfortunately this representation has infinitely many states. A finite representation can be obtained by truncating the infinite model. It is proven in [11] that an appropriate truncation at level k is able to preserve the joint distribution of the departure process up to lag-$(k-1)$. The truncation at level k is done as

$$D_0^{(k)} = \begin{bmatrix} L_0 & F & & \\ & L & F & \\ & & \ddots & \ddots \\ & & & L+F \end{bmatrix} \begin{matrix} 0 \\ 1 \\ \vdots \\ k \end{matrix}, \quad D_1^{(k)} = \begin{bmatrix} B & & \\ & \ddots & \\ & & B-FG & FG \end{bmatrix} \begin{matrix} 0 \\ 1 \\ \vdots \\ k \end{matrix}, \tag{31}$$

where matrix G is the minimal non-negative solution to the matrix-quadratic equation $0 = B + LG + FG^2$.

Although the truncation leads to a finite model, the number of states can still be too large. The superposition operations in the queueing network increase the number of states even more, and the limits of numerical tractability are easily hit. A possible solution for the state-space explosion is provided in [4], where a compact representation is constructed while maintaining the lag-1 joint moments of the large process.

5.2 Practical Problems and Possible Solutions

An issue with both the ETAQA departure model and the joint moment based approach is that they do not always return a Markovian representation, it is not even guaranteed that the departure model is a valid stochastic process.

Applying the RAP approximation procedure presented in Section 4 makes it possible to overcome this problem. Based on $(D_0{}^{(k)}, D_1{}^{(k)})$ it always returns a valid Markovian representation (H_0, H_1), and at the same time it is also able to compress the truncated departure process to a desired level.

There is, however, one issue which has to be taken account when applying the procedure of Section 4, namely that the number of marginal moments that can be used to obtain matrix H_0 is limited. We are going to show that the order of the PH distribution representing the inter-departure times is finite (denoted by N_D), determined by $2N_D - 1$ moments, and using more moments during the approximation leads to a dependent moment set (see [3]).

Theorem 5. *The order of the PH distribution representing the inter-departure times of a QBD queue with block size $N > 1$ is*

$$N_D = 2N. \tag{32}$$

Proof. In [11] it is shown how an order $2N$ PH distribution is constructed that captures the inter-departure times in an exact way, thus $N_D \leq 2N$. Additionally, it is easy to find concrete matrices B, F, L and L_0 such that the order of this PH distribution is exactly $2N$ (practically any random matrices are suitable, the order can be determined by the STAIRCASE algorithm of [2]). Consequently, we have that $N_D = 2N$. $\square$

Surprisingly, in case of MAP/MAP/1 queues the order of the inter-departure times is lower.

Theorem 6. *([4], Theorem 2) The order of the PH distribution representing the inter-departure times of a MAP/MAP/1 queue is*

$$N_D = N_A + N_S, \tag{33}$$

where N_A denotes the size of the MAP describing the arrival process and N_S the one of the service process, assuming that $N_A + N_S > 1$.

Thus, the proposed method for producing a MAP (B_0, B_1) that approximates the departure process is as follows:

1. First the ETAQA departure model is constructed up to the desired lag k, providing matrices $(D_0^{(k)}, D_1^{(k)})$. The stationary phase distribution at departure instans needs to be determined as well, α_D is the unique solution to $\alpha_D(-D_0^{(k)})^{-1}D_1^{(k)}, \alpha_D \mathbb{1} = 1$.
2. The marginal moments of the inter-departure times are computed from α_D and $D_0^{(k)}$. The more moments are taken into account, the larger the output of the approximation is. According to the above theorems, more than $2N_D - 1$ should not be used.
3. Matrix B_0 is obtained by moment matching (see Section 4.2).
4. Matrix B_1 is obtained such that either the squared distance of the joint density is minimized up to lag k, see 4.1.

5.3   Numerical Example

In this example[2] we consider a simple tandem queueing network of two MAP/MAP/1 queues. The arrival process of the first station is given by matrices

$$D_0 = \begin{bmatrix} -0.542 & 0.003 & 0 \\ 0.04 & -0.23 & 0.01 \\ 0 & 0.001 & -2.269 \end{bmatrix}, \quad D_1 = \begin{bmatrix} 0.021 & 0 & 0.518 \\ 0 & 0.17 & 0.01 \\ 0.004 & 0.005 & 2.259 \end{bmatrix}, \quad (34)$$

while the matrices characterizing the service process are

$$S_0 = \begin{bmatrix} -10 & 0 \\ 0 & -2.22 \end{bmatrix}, \quad S_1 = \begin{bmatrix} 7.5 & 2.5 \\ 0.4 & 1.82 \end{bmatrix}. \quad (35)$$

With these parameters both the arrival and the service times are positively correlated ($\rho_1^{(A)} = 0.21$ and $\rho_1^{(S)} = 0.112$) and the utilization of the first queue is 0.624.

The service times of the second station are Erlang distributed with order 2 and intensity parameter 6 leading to utilization 0.685.

This queueing network is analyzed such a way, that the departure process is approximated by the ETAQA truncation and by the joint moments based methods. Next, our RAP approximation procedure (Section 4) is applied to address the issues of the approximate departure processes, namely to obtain a Markovian approximation and in case of the ETAQA truncation method, to compress the large model to a compact one.

Table 1 depicts the mean queue length of the second station and the model size by various departure process approximations. The ETAQA truncation model has been applied with truncation levels 2 and 6, which has been compressed by our method based on either 3 or 5 marginal moments and with $\mathcal{D}_2\{\}$ or

---

[2] The implementation of the presented method and all the numerical examples can be downloaded from http://www.hit.bme.hu/~ghorvath/software

Table 1. Results of the queueing network example

Model of the departure process	#states	E(queue len.)
Accurate result (simulation):	n/a	2.6592
ETAQA, lag-1 truncation	18	2.3379
Our method based on 3 moments and $\mathcal{D}_2\{\}$	2	2.4266
Our method based on 5 moments and $\mathcal{D}_2\{\}$	3	2.5722
ETAQA, lag-5 truncation	42	2.5405
Our method based on 3 moments and $\mathcal{D}_2\{\}$	2	2.4266
Our method based on 5 moments and $\mathcal{D}_2\{\}$	3	2.5722
Our method based on 3 moments and $\mathcal{D}_6\{\}$	2	2.4266
Our method based on 5 moments and $\mathcal{D}_6\{\}$	3	2.6805
Joint moments based, 2 states	2	2.3255
Our method based on 3 moments and $\mathcal{D}_2\{\}$	2	2.3255
Joint moments based, 3 states	3	2.755
Our method based on 3 moments and $\mathcal{D}_2\{\}$	2	2.4266
Our method based on 5 moments and $\mathcal{D}_2\{\}$	3	2.7489

$\mathcal{D}_6\{\}$ distance optimization. The corresponding queue length distributions at the second station are compared in Figure 4. The departure process has also been approximated by the joint moments based method of [4], and an approximate Markovian representation has been constructed with our method based on 3 or 5 marginal moments and $\mathcal{D}_2\{\}$ optimization.

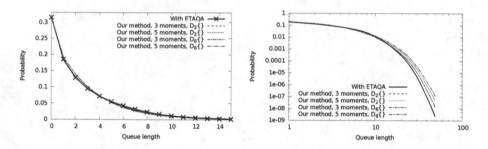

Fig. 4. Queue length distribution with the ETAQA departure model and its Markovian approximations

The results indicate that the RAP approximation and state space compression technique presented in this paper is efficient, the MAP returned is able to capture the important characteristic of the target RAP with an acceptable error.

Acknowledgments. This work was supported by the Hungarian research project OTKA K101150 and by the János Bolyai Research Scholarship of the Hungarian Academy of Sciences.

References

1. Asmussen, S., Koole, G.: Marked point processes as limits of Markovian arrival streams. Journal of Applied Probability, 365–372 (1993)
2. Buchholz, P., Telek, M.: On minimal representations of rational arrival processes. Annals of Operations Research **202**(1), 35–58 (2013)
3. Casale, G., Zhang, E.Z., Smirni, E.: Trace data characterization and fitting for Markov modeling. Performance Evaluation **67**(2), 61–79 (2010)
4. Horváth, A., Horváth, G., Telek, M.: A joint moments based analysis of networks of MAP/MAP/1 queues. Performance Evaluation **67**(9), 759–778 (2010)
5. Horváth, G.: Moment matching-based distribution fitting with generalized hyper-erlang distributions. In: Dudin, A., De Turck, K. (eds.) ASMTA 2013. LNCS, vol. 7984, pp. 232–246. Springer, Heidelberg (2013)
6. Johnson, M.A., Taaffe, M.R.: Matching moments to phase distributions: Mixtures of Erlang distributions of common order. Stochastic Models **5**(4), 711–743 (1989)
7. Latouche, G., Ramaswami, V.: Introduction to matrix analytic methods in stochastic modeling, vol.5. Society for Industrial and Applied Mathematics (1987)
8. Lipsky, L.: Queueing Theory: A linear algebraic approach. Springer Science & Business Media (2008)
9. Steeb, W.: Matrix calculus and Kronecker product with applications and C++ programs. World Scientific (1997)
10. Telek, M., Horváth, G.: A minimal representation of Markov arrival processes and a moments matching method. Performance Evaluation **64**(9), 1153–1168 (2007)
11. Zhang, Q., Heindl, A., Smirni, E.: Characterizing the BMAP/MAP/1 departure process via the ETAQA truncation. Stochastic Models **21**(2–3), 821–846 (2005)

Task Delegation in a Peer-to-Peer Volunteer Computing Platform

Kristóf Attila Horváth[1] and Miklós Telek[1,2]($\boxtimes$)

[1] Budapest University of Technology and Economics, Budapest, Hungary
horvath.kristof.attila@gmail.com
[2] MTA-BME Information systems research group, Budapest, Hungary
telek@webspn.hit.bme.hu

Abstract. The paper reports an effort made for understanding the effect of task delegation policy in a peer-to-peer volunteer computing platform. This effort includes the implementation of a simulation environment and the development of associated analytical models for the analysis of task delegation policies in peer-to-peer computing platforms. Based on the analytical model best and worst task delegation policies are computed and the resulted system behavior is verified by simulation.

Keywords: Peer-to-peer volunteer computing platform · Task delegation · Mean field model · Simulation

1 Introduction

The concept of utilizing the unutilized computing resources of a large number of (personal) computers connected via the internet is around for several decades. There are widely known peer-to-peer volunteer computing platform projects established for evaluating various computationally intensive tasks (a summary is provided in the next section). The related literature discusses the introduction, the spread, the order of magnitude, the organization and the applied technical details of these projects. In this work we focus on a particular detail of the organization of peer-to-peer volunteer computing platforms, the subtask delegation policy.

As the organization of volunteer computing platform changes from centrally controlled to peer-to-peer based, by time it became important to understand the performance consequences of autonomous subtask delegation policies.

The rest of the paper is organized as follows. Section 2 introduces the existing computing platforms and the related literature. We summarize the main properties of a proposed peer-to-peer volunteer computing platform in Section 3. Analytical models and associated performance analysis of various parameters of interests are investigated in Section 4. Finally, Section 5 presents the simulation results and their relation to the results of the analytical models.

The authors thank the support of the OTKA K101150 project.

M. Gribaudo et al. (Eds.): ASMTA 2015, LNCS 9081, pp. 115–129, 2015.
DOI: 10.1007/978-3-319-18579-8_9

2   Existing Volunteer Computing Platform Solutions

2.1   A Brief History of Volunteer Computing

There is a huge amount of unused computing capacity in personal computers, because the computers of users work 100% occupancy only negligible part of the time. This was the basis of the volunteer computing networks which utilize the unused capacity of personal computers. A study was published about the capacity of volunteer computing networks in 2006 [4], despite of old data the measured values are shocking: an ordinary volunteer computing project could use 95.5 teraFLOPS (10^{12} Floating-point Operations Per Second) computing capacity and 7.74 petabyte (10^{15} byte) storage.

The first volunteer computing projects started in 1997: the GIMPS (Great Internet Mersenne Prime Search) and the Distributed.net where cryptographic algorithms were tested. These projects had got tens of thousands of volunteer users [2].

The first project, which already had got millions of volunteer users, is the SETI@home project. It has started in 1999. SETI is the abbreviation of Search for Extraterrestrial Intelligence, and the @home (at home) suffix refers the use of personal computers instead of supercomputers. Tiny pieces of received signals of radio telescopes were sent to the computers of volunteers where the client application tried to find very narrowband (<Hz) signals in them. The method assumes that the extraterrestrial intelligence transmits narrowband signal which is easily distinguishable from the natural background radiation. The task is highly computationally intensive because a lot of parameters – bandwidth, symbol duration time, Doppler shift, etc. – are unknown [6, 13].

Because of the popularity of SETI@home project a general platform called BOINC (Berkeley Open Infrastructure for Network Computing) was developed in 2002. The BOINC platform became dominant in the subsequent years.

In the volunteer computing projects one of the hardest challenge is finding and keeping members as volunteers. Spectacular figures of the scientific results in wallpaper or in screen saver try to increase the interest. An other option is to publish the list of most effective volunteers and [17] recommends worker teams to utilize the team spirit. In spite of the seemingly infinite resources the performance optimization of distributed computing platforms is an essential goal [1].

2.2   Platforms

BOINC was developed at the University of California, Berkeley. It is the largest volunteer computing platform so far. The projects of the platform are computed on 600 000 personal computers. The total computing capacity almost reaches 10 petaFLOPS, therefore the system rivals the most powerful supercomputers [19]. Apart of the SETI@home project the platform hosts additional projects like Einstein@home, LHC@home, Milkyway@home, etc [3, 5].

The XtremWeb platform was developed in parallel and independently from the BOINC system. The objective and the implementation are very similar in the two platforms, however in the competition for users BOINC was more successful [11].

Alchemi is a .NET-based platform, which was developed at the University of Melbourne. In this system the main objective is the easy programmability, the other aspects are less important [15,16].

The OurGrid platform is based on a new idea. This platform interconnects the grid systems of universities and research groups intend to utilize the free resources [7,8].

All of these platforms follow the master–worker parallel programming paradigm. The central server decomposes the task into subtasks and manages the delegation. The lifecycle of a subtask is the following: (a) the server creates a job by packing the executable code and the input files together (b) the client downloads the job (c) the client computes the results (d) the client uploads the results (e) the server verifies and processes the results.

3   Properties of the Proposed Distributed Computing Platform Solution

The existing volunteer computing platforms have got two main issues. The first one is the protection of the volunteer's computer. In BOINC and XtremWeb the servers send native executable code to the clients. The platforms use asymmetric cryptography to ensure the authenticity and the integrity, but the project owner can execute anything on the volunteer's computer.

On the other hand Alchemi and OurGrid systems use virtualization to solve the problem, but it reduces the performance which should be avoided in a computing platform. In the proposed computing platform the elements send the source code to the peers. This method includes filtering of malicious codes. The compiling–running combination may be more efficient than the virtualization.

The second issue is that the management of subtasks is centralized in all existing platforms; they follow the master–worker programming paradigm. In some systems the executable code or input files can be shared by peer-to-peer mechanisms [9,10], however the central management of subtasks are presented here as well.

In the proposed distributed volunteer computing platform, every node can delegate subtasks to other nodes, if the currently computed subtask contains parallel blocks. In this approach the programmer can write multi-level subtask structures, so a subtask in a parallel block can be the same as the main program following a fractal-like structure. The nodes in the computing network are identical, so a homogenous programming model can be used instead of a heterogenous programming model as in CUDA [12], OpenCL [18], etc.

The codes can contain serial- and parallel blocks sequentially. The parallel blocks must be fully decomposable, so the subtasks can not communicate with each other or with the main task (except the interchange through input- and output files). The next serial- or parallel block can be started after all subtasks of the current block had been completed.

At the beginning of the parallel blocks the client program decides how many subtasks will be computed locally and how many will be delegated. Because of the overhead of subtask delegation there is a trade-off between locally computed and delegated subtasks. The optimal strategy is investigated and the behavior of the whole system is analysed in the following section.

4   Performance Analysis of Distributed Computing Platforms

In this section we investigate the performance of distributed computing platforms with different analysis approaches. The two main applied analytical approaches are the phase type (PH) distributions and the mean field approximations, which we also summarize below.

4.1   Phase Type Distributions

If the stochastic behavior of a real system can be characterized by a Markov chain then various random event times which are of practical interests are PH distributed. This statement applies for both discrete and continuous time Markov chains (DTMCs and CTMCs) with associated discrete or continuous PH distributions. In this work we focus on continuous time models.

Definition 1. *The time to reach the absorbing state in a CTMC with n transient and an absorbing state is (size n) phase type distributed.*

Consequently, a (continuous) PH distributed random variable $\mathcal{X}$ is continuous non-negative with cumulative distribution function

$$F(t) = Pr(\mathcal{X} < t) = 1 - \underline{v}e^{\mathbf{H}t}\mathbb{1} ,$$

where row vector $\underline{v}$ contains the initial probabilities of the CTMC in the transient states, square matrix $\mathbf{H}$ contains the transition rates among the transient states and column vector $\mathbb{1}$ is composed by ones. $\underline{v}$, $\mathbf{H}$ and $\mathbb{1}$ are referred to as initial probability vector, transient generator matrix and closing vector, respectively. Throughout the paper we assume that the Markov chain starts from a transient state, i.e., $\underline{v}\mathbb{1} = 1$, and consequently $\mathcal{X}$ has no probability mass at zero. The density and the moments of $\mathcal{X}$ are

$$f(t) = \underline{v}e^{\mathbf{H}t}(-\mathbf{H})\mathbb{1} , \tag{1}$$

$$\mu_n = E(\mathcal{X}^n) = n!\underline{v}(-\mathbf{H})^{-n}\mathbb{1} . \tag{2}$$

If the initial probability vector and the transient generator matrix are obtained from modeling assumptions all performance measures associated with $\mathcal{X}$ can be computed based on (1) and (2).

4.2   Execution Time Model

The execution time of a parallel block can be analyzed by the CTMC shown in Figure 1. The states can be arranged in a two-dimensional grid: horizontal axis shows the number of locally computed subtasks, the vertical axis shows the number of delegated subtasks, which are computed on other computers.

The absorbing state, which represents the completion of a task, is state $(0,0)$. The CTMC in Figure 1 describes the case when a subtask can be computed locally in an exponentially distributed amount of time with parameter μ_1, and a delegated subtask can be sent, computed and returned in an exponentially distributed amount of time with parameter μ_2.

From any states (except the states at the top and the left boundaries) there are two possible transitions:

1. a locally-computed subtask completes with rate μ_1 and the process goes from state (i,j) to state $(i,j-1)$,
2. one of the i delegated subtasks arrives with rate $i\mu_2$ and the process goes from state (i,j) to state $(i-1,j)$.

There is no local (dedicated) computation at the left (top) boundary, so there is only one of the two transitions at these boundaries.

Based on the transition graph the infinitesimal generator matrix of the process $\mathbf{Q}$ is obtained by mapping the nodes of the transition graph, which are the states of the Markov chain, to a subset of natural numbers and indicating the transition rates between the pair of states. The transient generator, matrix $\mathbf{H}$ in (1) and (2), contains the transition rates only among the transient states, and is the lower right block of matrix $\mathbf{Q}$ as it is indicated below for the state space with 3×3 states.

$$
\mathbf{Q} = \begin{bmatrix}
0 & 0 & 0 & 0 & 0 & 0 & 0 & 0 & 0 \\
\mu_1 & -\mu_1 & 0 & 0 & 0 & 0 & 0 & 0 & 0 \\
0 & \mu_1 & -\mu_1 & 0 & 0 & 0 & 0 & 0 & 0 \\
\mu_2 & 0 & 0 & -\mu_2 & 0 & 0 & 0 & 0 & 0 \\
0 & \mu_2 & 0 & \mu_1 & -\mu_1-\mu_2 & 0 & 0 & 0 & 0 \\
0 & 0 & \mu_2 & 0 & \mu_1 & -\mu_1-\mu_2 & 0 & 0 & 0 \\
0 & 0 & 0 & 2\mu_2 & 0 & 0 & -2\mu_2 & 0 & 0 \\
0 & 0 & 0 & 0 & 2\mu_2 & 0 & \mu_1 & -\mu_1-2\mu_2 & 0 \\
0 & 0 & 0 & 0 & 0 & 2\mu_2 & 0 & \mu_1 & -\mu_1-2\mu_2
\end{bmatrix} \tag{3}
$$

The elements under the diagonal contain μ_1 except every i^{th}, which is 0, because local computing does not happen at the first column of the grid. The ith elements under the diagonal contain the rates of arriving delegated task, which are μ_2, $2\mu_2$, etc. in i-length blocks. Conventionally, the diagonal elements contain the negation of sum of all the other elements in the row.

Expected Value of the Task Execution Time. The mean task execution time (the running time of parallel block), which is the mean time to reach state

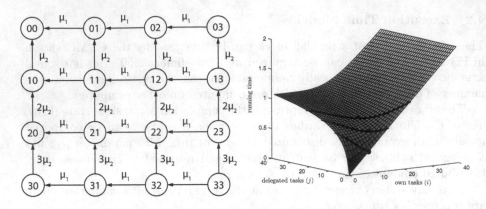

Fig. 1. Phase type CTMC for execution   **Fig. 2.** Running times for different states
time modeling

$(0,0)$ in the Markov chain in Figure 1, is an essential performance measure of
the distributed computing platform in order to optimize the delegation strategy.
In general, the mean task execution time can be computed based on (2), but
utilizing the structural properties of matrix $\mathbf{H}$ it can also be computed by an
efficient recursive procedure.

If the process is in state (i,j), $i > 0$, $j > 0$, it can move forward to one of the
two possible consecutive states. On both cases we can compute the probability of
taking one of the two possible consecutive states and the expected time to reach
the $(0,0)$ state from the next state. If the local calculation completes sooner and
the process goes on the first trajectory we have:

$$E(\hat{T}_{i,j}) = \frac{1}{\mu_1 + i\mu_2} + E(T_{i,j-1}) . \tag{4}$$

If one of the delegated tasks completes sooner and the process goes on the second
trajectory:

$$E(\check{T}_{i,j}) = \frac{1}{\mu_1 + i\mu_2} + E(T_{i-1,j}) . \tag{5}$$

To calculate the full time, the two conditional expected times, (4) and (5), have
to be weighted by the probabilities of the trajectories.

$$E(T_{i,j}) = \Pr(1^{\text{st}} \text{ trajectory})E(\hat{T}_{i,j}) + \Pr(2^{\text{nd}} \text{ trajectory})E(\check{T}_{i,j}) \tag{6}$$

$$= \frac{\mu_1}{\mu_1 + i\mu_2}\left(\frac{1}{\mu_1 + i\mu_2} + E(T_{i,j-1})\right) + \frac{i\mu_2}{\mu_1 + i\mu_2}\left(\frac{1}{\mu_1 + i\mu_2} + E(T_{i-1,j})\right) .$$

This recursive formula is valid for every state where $i > 0$ and $j > 0$. The mean
task execution time from the boundary states where $i = 0$ or $j = 0$ can be
computed as follows. If the phase type distribution is initialized from a state
where $i = 0, j > 0$, the task execution time is Erlang(j, μ_1) distributed (the sum

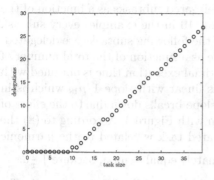
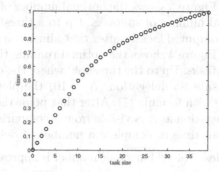

Fig. 3. Optimal number of delegated sub-tasks

Fig. 4. Optimized running time

of j independent exponentially distributed random times with the parameter μ_1), whose expected time is

$$E(T_{0,j}) = j\frac{1}{\mu_1}. \tag{7}$$

In case of $j = 0, i > 0$ the successive exponential phases do not have the same parameter, so the expected values of the phases with phase dependent rates ($k\mu_2$, $k = i, i-1, \ldots, 1$) have to be summed up.

$$E(T_{i,0}) = \sum_{k=1}^{i} \frac{1}{k\mu_2} = \frac{1}{\mu_2} \sum_{k=1}^{i} \frac{1}{k} = \frac{1}{\mu_2} H_i, \tag{8}$$

where $H_i = \sum_{k=1}^{i} \frac{1}{k}$ is commonly referred to as the ith harmonic number.

Optimization. To investigate the optimal subtask delegation we evaluate the mean task execution time from different initial states. Figure 2 depicts the mean task execution time as a function of the initial state, when $\mu_1 = 20$ and $\mu_2 = 4$. Indeed the continuous surface on Figure 2 is valid only at integer points.

If the overhead of delegation reduces and the ratio of two intensities converges to zero, then the surface would converge to the $E(T) = i\mu_1$ plane.

When the program reaches a parallel block, it has to decide how many subtasks will be computed locally and how many will be delegated. This two numbers determine the starting state of the process. When a parallel block is composed by N subtasks the starting state satisfies the $i + j = N$ formula, therefore the optimum is the minimum of the surface on the $i + j = N$ section plane. The section plane has got N integer points, the optimal running time is obtained at the minimum of these. Figure 2 contains the $N = i + j = 8, 15, 23$ section plane on the surface and their minimum points. These points indicate the optimal number of delegated subtask in case of n parallel tasks.

Figure 3 shows the optimal number of delegated subtasks as a function of the total number of subtasks. Up to a threshold (10 in the example) every subtask is computed locally, after that almost all of the following subtasks are delegated.

Figure 4 shows the optimal running time as a function of the total number of subtasks. Up to the threshold where the optimal execution time is obtained with no subtask delegation ($N < 10$) the plot is linear with slope $1/\mu_1$, which is in line with formula (7). After that point the slope breaks down due to the effect of delegation as it is visible from a comparison with Figure 3. According to (8) the mean time to complete a number of delegated task is related to the harmonic series. The ith harmonic number is approximately equal to $\ln(i)$, because $\int_1^n \frac{1}{x} dx = \ln(n)$ and the harmonic series is an approximation to the definite integral. The difference between the harmonic series and the logarithmic series converges to the Euler–Mascheroni constant: $\lim_{n\to\infty} H_n - \ln(n) = \gamma \approx 0.5772156649$.

Model Parameters Setting. The expected value of running time of a locally-computed subtask is the reciprocal of μ_1, so μ_1 can be defined as the FLOPS of the machine divided by the floating-point operations of a subtask. Similarly, the reciprocal of μ_2 can be defined as the sum of the floating-point operations of one subtask divided by the average FLOPS in the computing network and the mean time to transmit the input/output data through the internet connection.

The above described Markovian model and this parameters setting do not guarantee the perfect matching of the model and the real execution times in the computing platform. For this reason we also verify the results by simulation in Section 5.

4.3   Mean Field Approximation

In the previous subsections we optimized the system behavior assuming infinite computing resources. To consider the effect of finite computing resources we apply a different modeling approach, the mean field approximation. The mean field method allows the analysis of large Markov systems which are composed by a finite number of identical interacting components, where the interaction depends only on the number of components which are in particular states. For example, in our case the identical components are the computing units and the dependence of one component on the other components is only through the number of available idle computing units. Let N be the number of components. The state of component ℓ ($\ell = 1, 2, \ldots, N$) at time t is denoted by $X_\ell(t)$. In our case the state of the component depends on the task it is working on. A component could be idle, working on a delegated task, and being responsible for the execution of the task composed by parallel subtasks. The latest case can be described by a state space similar to the one on Figure 1. The overall behavior of these three possible cases is discussed below and depicted on Figure 5. The state space of each component, S, is composed by $s = |S|$ states, and $N_i(t)$ denotes the number of components which are in state i ($\forall i \in S$) at time t. For example,

$N_{idle}(t)$, denotes the number of idle computing units which are available for task or subtask assignment. In our case it is intuitive to see that a delegation decision depends only on the number of idle computing units and not on the state of a particular computing unit of the system. The row vector composed by $N_i(t)$ is denoted by $\mathbf{N}(t)$ and by this definition, $\sum_{i=1}^{s} N_i(t) = N$ (is the number of all computing units in our case).

The global behavior of the set of N components forms a CTMC over the state space of size s^N. However, due to the fact that the components are identical and indistinguishable, the state space can be lumped into the aggregate state space S_L of size $\binom{N+s-1}{s-1}$, where a state of the overall CTMC is identified by the number of components staying in each state of S, i.e., by $\mathbf{N}(t) = (N_1(t), N_2(t), \ldots, N_s(t))$ (in our case it means the number of idle components, the number of components working on a delegated task, the number of components responsible for the computation of a task and waits for the completion of two delegated subtasks, etc). $\mathbf{N}(t)$ refers to the population vector, which describes the distribution of the population between the possible states.

The evolution of a given computing unit is such that the transition rates may depend on the global behavior through the actual value of vector $\mathbf{N}(t)$. With this assumption, the transition rate of a particular component from state i to j is $K_{ij}(\mathbf{N}(t))$ which depends only on vector $\mathbf{N}(t)$. The diagonal elements of the transition rate matrix are defined by $K_{ii}(\mathbf{N}(t)) = - \sum_{j \in S, j \neq i} K_{ij}(\mathbf{N}(t))$.

Instead of using the population vector, $\mathbf{N}(t)$, the normalized population vector, $\mathbf{n}(t) = \mathbf{N}(t)/N$ is commonly used for the mean field analysis of such systems. The entries of $\mathbf{n}(t)$ define the proportion of objects in state i at time t and $\sum_{i \in S} n_i(t) = 1$. The associated transition rate function is denoted by $k_{ij}(\mathbf{n}(t))$, and the matrix composed by these elements is $\mathbf{k}(\mathbf{n}(t)) = \{k_{ij}(\mathbf{n}(t))\}$. Hereafter we assume that $k_{ij}(\mathbf{n}(t))$ is a Lipschitz continuous function over the s-dimensional unit cube. The mean field method is based on the following essential theorem.

Theorem 1. *[14] The normalized state vector of the population process, $\mathbf{n}(t)$, tends to be deterministic, in distribution, as N tends to infinity and satisfies the following differential equation*

$$\frac{d}{dt}\mathbf{n}(t) = \mathbf{n}(t)\,\mathbf{k}(\mathbf{n}(t))\,. \tag{9}$$

The mean field approximation is based on the fact that for large but finite N the deterministic approximation according to (9) is a good approximation of the system behavior with well defined error bounds [14].

4.4   The Mean Field System Model

The model in Section 4.2 optimizes the task execution time disregarding its effect on resource utilization. In this section, we present an approximate mean field model of the system which considers also the resource utilization.

When the studied volunteer computing project runs a task with one level delegation policy (delegated subtasks are not divided into lower level subsubtasks), then the state space of a computing device is similar to the state space of the execution time model when the computing device is responsible for the computation of a task composed of a given number of subtasks. In this case state (i, j) is the state when the computing device is waiting for the completion of i delegated subtasks and the node has to compute j subtasks locally. In state $(0, 0)$ the node is idle, so it can start with a new task or it can receive a delegated subtask from another node. Additional to the states associated with the task computation there is state $(*)$ which identifies the state when the node computes a delegated subtask.

To describe the considered subtask delegation policy we introduce the delegation function f_{ij} as follows

$$f_{ij} = \Pr(\text{the node goes to state } (i, j) \text{ after the delegation}). \tag{10}$$

For notational convenience we assume $f_{00} = 0$. $\sum_i \sum_j f_{ij} = 1$, because f_{ij} describes the probability of a complete and disjoint set of events. Indeed, this delegation function also contains the distribution of the number of parallel subtasks in an arriving main task.

$$\Pr(\text{an arriving task is composed by } k \text{ parallel subtasks}) = \sum_{i=0}^{k} f_{i,k-i}. \tag{11}$$

Further more, λ denotes the arrival intensity of a new task. A new task is composed by a given number of parallel subtasks and, similarly to the notations of the previous sections, μ_1 and μ_2 denote the subtask completion rate for local and delegated subtasks, respectively. In the $i \leq 3$, $j \leq 3$ part of the state space the following differential equations describe the evolution of the normalized population vector.

$$\frac{d}{dt}n_{01}(t) = -\mu_1 n_{01}(t) + \mu_2 n_{11}(t) + \mu_1 n_{02}(t) + \lambda f_{01},$$

$$\frac{d}{dt}n_{02}(t) = -\mu_1 n_{02}(t) + \mu_2 n_{12}(t) + \mu_1 n_{03}(t) + \lambda f_{02},$$

$$\frac{d}{dt}n_{03}(t) = -\mu_1 n_{03}(t) + \mu_2 n_{13}(t) + \lambda f_{03},$$

$$\frac{d}{dt}n_{10}(t) = -\mu_2 n_{10}(t) + 2\mu_2 n_{20}(t) + \mu_1 n_{11}(t) + \lambda f_{10},$$

$$\frac{d}{dt}n_{11}(t) = -\mu_2 n_{11}(t) - \mu_1 n_{11}(t) + 2\mu_2 n_{21}(t) + \mu_1 n_{12}(t) + \lambda f_{11},$$

$$\vdots$$

The general form of these equations is

$$\frac{d}{dt}n_{ij}(t) = -(i\mu_2 + \mu_1)n_{ij}(t) + \mathcal{I}(i+1)\mu_2 n_{i+1,j}(t) + \mathcal{I}(j+1)\mu_1 n_{i,j+1}(t) + \lambda f_{ij},$$
(12)

where

$$\mathcal{I}(i) = \begin{cases} 1, & \text{if } i \leq 3, \\ 0, & \text{othervise.} \end{cases}$$

The number of delegated subtasks from one main task is $\sum_i i \sum_j f_{ij}$, consequently the same number of nodes move to state $(*)$ at the arrival of a main task due to the associated task delegations. The differential equation for the number of nodes in state $(0,0)$ is

$$\frac{d}{dt}n_{00}(t) = -\lambda\left(1 + \sum_i i \sum_j f_{ij}\right) + \mu_1 n_{01}(t) + \mu_2 n_{10}(t) + \mu_2 n_*(t) .$$
(13)

Finally, the differential equation for state $(*)$ is

$$\frac{d}{dt}n_*(t) = \lambda \sum_i i \sum_j f_{ij} - \mu_2 n_*(t) .$$
(14)

In this set of differential equations the number of computing units working on a delegated subtask is encoded in two ways

$$n_*(t) = \sum_i i \sum_j n_{ij}(t).$$
(15)

From the fact that the normalized population vector sums up to one we further have

$$\sum_i \sum_j n_{ij}(t) + n_*(t) = 1.$$
(16)

These two relations can be used to simplify the system description or for sanity check of results obtained from the redundant description.

The state transitions described by the differential equations are illustrated in Figure 5. The vertices represent the possible states of the computing units, the edges represent the state transitions and the associated intensities indicate the transition rates. Note that in some cases intensities depend on some relative population values (which is not the case with a transition graph of a CTMC). Those are the cases when the $\mathbf{k}(\mathbf{n}(t))$ matrix elements depend on $\mathbf{n}(t)$ as it is in (9). This set of differential equations can be solved by numerical procedures, e.g., by Runge–Kutta method.

Starting from a completely idle system, $\mathbf{n}(0) = (1, 0, \cdots, 0)$, Figure 6 depicts the system evolution when $\lambda = 1$, $\mu_1 = 5$, $\mu_2 = 4$ and $f_{21} = 1$. According to Figure 6 the system is not saturated with this load and the normalized population vector converges to a fix point.

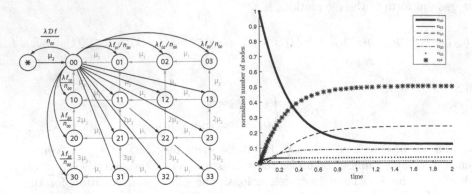

Fig. 5. Transition graph of the mean field model   **Fig. 6.** The results of mean field analysis

The analysis of the behavior of an overloaded system requires an extension of the above set of differential equations with boundary limits to keep the normalized population values between 0 and 1. For example the boundary extension of (13) is

$$\frac{d}{dt}n_{00}(t) = \begin{cases} \bullet, & \text{if } 0 < n_{00}(t) < 1, \\ \max(\bullet, 0), & \text{if } n_{00}(t) = 0, \\ \min(\bullet, 0), & \text{if } n_{00}(t) = 1, \end{cases} \tag{17}$$

where $\bullet$ stands for the expression on the right hand side of (13). The limit of stability is the highest load where $\lim_{t\to\infty} n_{00}(t) > 0$ still holds.

Stability with Autonomous Computing Units

In volunteer computing platforms the project owner has to avoid the system saturation. It is not a trivial problem when the participating computing units are autonomous. For example, if the delegation function is not known because the nodes decide the delegation strategy autonomously the project owner has to find a safe task submission rate, λ, at which the system remains stable independent of the delegation policy of the autonomous computing units.

A safe task submission rate can be obtained by assuming that the autonomous computing units apply always the most inefficient task delegation policy.

The overall resource utilization of the task completion with i delegated and j locally computed subtasks, C_{ij}, can be computed by a recursive relation based on the same considerations as the ones in Section 4.2

$$E(C_{i,j}) = \frac{\mu_1}{\mu_1 + i\mu_2}\left(\frac{i+1}{\mu_1 + i\mu_2} + E(C_{i,j-1})\right) + \frac{i\mu_2}{\mu_1 + i\mu_2}\left(\frac{i+1}{\mu_1 + i\mu_2} + E(C_{i-1,j})\right),$$

which accounts for the total resource utilization, since with i delegated and j locally computed subtasks $i+1$ computing units are occupied. Plotting the obtained $E(C_{i,j})$ values similar to Figure 2 and taking the maximal values along

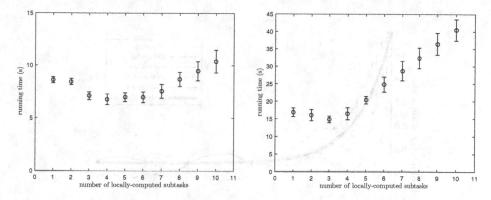

Fig. 7. The results of simulation with 10 subtasks

Fig. 8. The results of simulation with 20 subtasks

the $i + j = N$ section planes defines the most inefficient delegation policy. The limit of saturation with this most inefficient delegation policy defines the admissible safe task submission rate of the project owner.

We close the section by mentioning, that the modeling approach applied in this section can be extended for higher levels of hierarchical task subdivisions, but it would complicate the discussion significantly and is out of the scope of this paper.

5   Simulation

To verify the analytical results we have developed an event-driven simulator in C++ language. The implemented simulation model contains several additional details of a real volunteer computing platform including socket management for data transfer between nodes, task execution in virtual environment, realistic computation delegation strategy with higher level hierarchical subtask division, etc.

5.1   Execution Time Results

Figure 7 shows the execution times of a task with 10 subtasks. In this example, the subtasks can be computed with 10 MFLO (Mega FLoating-point Operations) and the computers have 10 MFLOPS computation capacity in average. The amount of data required for the computation of a subtasks is 3 MB, and the speed of the internet connection is 2 MB/s. Figure 7 shows the average running times and the confidence intervals for 95% confidence level.

The minimum of the task completion time is obtained at 4 locally-computed and 6 delegated subtasks. The running time is 6.76 s in this case.

Figure 8 shows another example with 20 subtasks. It contains the running times up to 10 locally computed subtasks because the linear trend continues

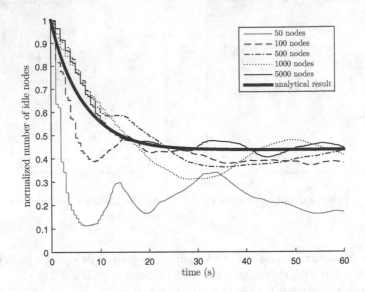

Fig. 9. The results of the simulation and the analytical model for relative number of nodes

over. In this example the subtasks can be computed with 40 MFLO, the other parameters are the same as above. The optimal delegation was found at 3 locally computed subtasks which coincides with the analytical results from Section 4.2. The associated optimal running time is 14.85 s.

5.2   Results of the Overall System Behavior

To verify the results of the mean field analysis in Section 4.3 we collected population results in the simulator. Figure 9 plots normalized number of idle nodes as a function of time for computing platforms with different number of computing units and identical relative loads. In this simulator run tasks were composed by 20 subtasks, the intensity of the task arrivals was $\lambda = 0.005$ 1/s, the calculated task completion rates were $\mu_1 = 0.27$ 1/s and $\mu_2 = 0.18$ 1/s. The figure contains results for computing platforms with 50, 100, 500, 1000 and 5000 nodes as well as the results of the mean field model. Figure 9 supports the intuition that the simulated results converge to the analytical result as the number of nodes increases.

6   Conclusions

The paper presents a performance assessment of volunteer computing platforms. A set of real characteristic features, e.g. task failures, arrivals and departures of computing nodes, restricted availability, local usage, etc. are left for future work. Different modeling paradigms are used for the analysis of performance measures.

References

1. Altman, E., Kameda, H., Hosokawa, Y.: Nash equilibria in load balancing in distributed computer systems. International Game Theory Review **4**, 91–100 (2002)
2. Anderson, D.P.: Public computing: reconnecting people to science. In: Conference on Shared Knowledge and the Web, pp. 17–19 (2003)
3. Anderson, D.P.: BOINC: a system for public-resource computing and storage. In: Fifth IEEE/ACM International Workshop on Grid Computing, pp. 4–10 (November 8, 2004)
4. Anderson, D.P., Fedak, G.: The computational and storage potential of volunteer computing. In: Sixth IEEE International Symposium on Cluster Computing and the Grid, CCGRID 2006, vol. 1, pp. 73–80. IEEE (2006)
5. Anderson, D.P., Walton, R., Fenton, C.: BOINC project. http://boinc.berkeley.edu/
6. Anderson, D.P., Werthimer, D.: SETI@home project. http://setiathome.berkeley.edu/
7. Andrade, N., Cirne, W., Brasileiro, F., Roisenberg, P.: OurGrid: an approach to easily assemble grids with equitable resource sharing. In: Feitelson, D.G., Rudolph, L., Schwiegelshohn, U. (eds.) JSSPP 2003. LNCS, vol. 2862, pp. 61–86. Springer, Heidelberg (2003)
8. Andrade, N., Costa, L., Germóglio, G., Cirne, W.: Peer-to-peer grid computing with the OurGrid community. In: Proceedings of the SBRC, pp. 1–8 (2005)
9. Costa, F., Silva, L., Fedak, G., Kelley, I.: Optimizing the data distribution layer of BOINC with BitTorrent. In: IEEE International Symposium on Parallel and Distributed Processing, IPDPS 2008, pp. 1–8. IEEE (2008)
10. Farkas, G., Szanto, I., Gora, V., Haller, P.: Extending the BOINC architecture using peer-to-peer application code exchange. In: 2011 Roedunet 10th International Conference (RoEduNet), pp. 1–4. IEEE (2011)
11. Fedak, G., Germain, C., Neri, V., Cappello, F.: Xtremweb: a generic global computing system. In: Proceedings of the First IEEE/ACM International Symposium on Cluster Computing and the Grid, pp. 582–587. IEEE (2001)
12. Garland, M.: Parallel computing with CUDA. In: Proc. of the IEEE Int. Symp. on Parallel and Distributed Processing (IPDPS), Atlanta, GA (April 19–23, 2010)
13. Korpela, E., Werthimer, D., Anderson, D.P., Cobb, J., Lebofsky, M.: SETI@home - massively distributed computing for SETI. Computing in science & engineering **3**(1), 78–83 (2001)
14. Kurtz, T.G.: Strong approximation theorems for density dependent Markov chains. Stochastic Processes and their Applications **6**(3), 223–240 (1978)
15. Luther, A., Buyya, R., Ranjan, R., Venugopal, S.: Alchemi: a. net-based grid computing framework and its integration into global grids. arXiv preprint cs/0402017 (2004)
16. Luther, A., Buyya, R., Ranjan, R., Venugopal, S.: Peer-to-peer grid computing and a. NET-based alchemi framework. High Performance Computing: Paradigm and Infrastructure. Wiley Press, Fall (2004)
17. Nov, O., Anderson, D., Arazy, O.: Volunteer computing: a model of the factors determining contribution to community-based scientific research. In: Proceedings of the 19th international conference on World wide web, pp. 741–750. ACM (2010)
18. Stone, J.E., Gohara, D., Shi, G.: Opencl: A parallel programming standard for heterogeneous computing systems. Computing in science & engineering **12**(3), 66 (2010)
19. Strohmaier, E., Dongarra, J., Simon, H., Meuer, M.: TOP500 project. http://www.top500.org/lists/2014/11/

On Convergence Rate to Stationarity of Queues with General Gaussian Input

Oleg Lukashenko[1,2]($\boxtimes$) and Evsey Morozov[1,2]

[1] Institute of Applied Mathematical Research of the Karelian Research Centre RAS, Petrozavodsk, Russia
lukashenko-oleg@mail.ru
[2] Petrozavodsk State University, Petrozavodsk, Russia
emorozov@karelia.ru

Abstract. The paper studies the rate of convergence to stationarity of the fluid queueing system with a constant service rate which is fed by a Gaussian process with stationary increments. It is assumed that variance of the input process is regularly varying with index $2H \in (1, 2)$. It is proved that the convergence rate is exactly the same that has been obtained for the fluid system fed by the corresponding fractional Brownian motion.

Keywords: Convergence · Stationarity · Fractional Brownian motion · Regular variation · Large deviations · Gaussian input

1 Introduction

Gaussian processes are well-recognized models to describe the traffic dynamics of a wide class of the modern telecommunication networks. The main motivation to apply these models is that the researchers have detected specific properties, such as self-similarity and long-range dependence, which are inherent in the modern network traffic [10,20]. Gaussian approximation is also motivated by statistical analysis of data traces [8,15]. We recall that self-similarity means that the distribution of the process remains unchanged under suitable scaling of time and space, while the long-range dependence means a slow decay of the autocorrelation function. These properties make difficult the probabilistic analysis and, as a consequence, the obtaining key characteristics in an explicit form. At the same time, these characteristics are crucial to evaluate the Quality of Service (QoS) provided by the networks.

We consider a centred Gaussian input process $A := \{A_t, \, t \in \mathbb{R}\}$ with stationary increments and with regularly varying variance,

$$\sigma^2(t) := \mathbb{V}\mathrm{ar} A_t = L(t)|t|^{2H}, \; 0 < H < 1, \tag{1}$$

E. Morozov—This work is supported by Russian Foundation for Basic research, projects 15–07–02341 A, 15–07–02354 A,15–07–02360 A, and also by the Program of strategic development of Petrozavodsk State University.

© Springer International Publishing Switzerland 2015
M. Gribaudo et al. (Eds.): ASMTA 2015, LNCS 9081, pp. 130–142, 2015.
DOI: 10.1007/978-3-319-18579-8_10

where L is a slowly varying function: $\lim_{t \to \infty} L(tx)/L(t) = 1$ for any $x > 0$. We mention the following important cases of Gaussian inputs.

1. *Fractional Brownian motion (fBm)*, denoted by B_H, satisfies (1) with $L \equiv 1$ and is the most studied self-similar long-range dependent Gaussian process. A fBm, being the input to a queueing system, is called *fractional Brownian (fB) input*. To motivate our interest to fB input, we consider N independent identically distributed (i.i.d.) *on-off sources*, such that source k is described by the process $\{I_k(t), \, t \geq 0\}$, $k = 1, ..., N$, with

$$I_k(t) = \begin{cases} 1, \, t \in \text{ on-period} \\ 0, \, t \in \text{ off-period}. \end{cases} \tag{2}$$

During an *on-period* a source is active, while it is inactive during the following *off-period*. The on-off periods form an *alternating renewal process*. The aggregated arrived workload generated by all sources during time interval $[0, t]$ is then defined as

$$A_N(t) = \int_0^t \left(\sum_{k=1}^N I_k(u) \right) du. \tag{3}$$

We assume that distribution function F of on-period (and/or off-period) is *heavy-tailed*, that is

$$1 - F(x) \sim cx^{-\alpha}L(x), \quad x \to \infty, \tag{4}$$

where $c > 0$ is a constant, parameter $\alpha \in (1, 2)$ and function L is slowly varying at infinity. Note that, because $\alpha > 1$ in (4), the mean periods $\mu_{on} < \infty$, $\mu_{off} < \infty$. It is shown in [19] that (under mild conditions) the following approximation holds for t and N large:

$$A(t) \approx tN \frac{\mu_{on}}{\mu_{on} + \mu_{off}} + \sqrt{L(t)N} B_H(t), \tag{5}$$

where slowly varying function L is expressed in the terms of given parameters, and the *Hurst* parameter $H \in (1/2, 1)$. This result allows to consider a queueing system fed by fB input as a suitable model for a wide class of the modern telecommunication systems.

2. *The sum of n independent fBms* which models the aggregated traffic generated by heavy-tailed *heterogeneous* on-off inputs with different parameters α_i satisfying (4), $i = 1, ..., n$. This superposition of on-off inputs, after an appropriate time scaling, converges weakly to the sum of independent fBms with variance $\sigma^2(t) = \sum_{i=1}^n t^{2H_i}$ and different parameters $H_i \in (1/2, 1)$ [19].

3. *Integrated Ornstein-Uhlenbeck process* with variance $\sigma^2(t) = t + e^{-t} - 1$, which is the Gaussian counterpart of the Anick-Mitra-Sondi fluid model [2] (see also [1]). The relevance of the latter model for the network traffic modelling is motivated in [9].

We assume (w.l.o.g.) that the service rate equals 1. According to [18] the stationary workload at instant t is defined as

$$Q(t) := \sup_{s \leq t}(A_t - A_s - (t - s)),\qquad\qquad (6)$$

and is distributed as [14]

$$M := \sup_{t \geq 0}(A_t - t).$$

An important performance measure of the communication systems is the *overflow probability* $\mathbb{P}(M > b)$ the stationary workload exceeds a finite threshold b. In an infinite buffer Gaussian system, analysis of the overflow probability is reduced to analysis of the extremes of Gaussian processes. For a queueing systems with general Gaussian input (including fB input), there are no explicit expression for the overflow probability, however a few asymptotic results are available. In this regard we mention the following key works [5–7]. It is important to emphasize that the asymptotic results proved below for the so-called *large buffer regime*, that is as $b \to \infty$, can be used for accurate approximation for moderate values of b as well. To determine the accuracy of the asymptotic approximation, it is reasonable to use simulation. In this regard it is quite important to know how long the simulation sample path should be, and this problem is closely relates to the rate of convergence to stationarity. Indeed, for each t, denote

$$M(t) := \sup_{s \in [0,t]}(A_s - s).\qquad\qquad (7)$$

Then the knowledge of the convergence rate allows to determine a time instant T such that, for $t \geq T$, an estimate of the overflow probability $\mathbb{P}(M(t) > x)$ approximates $\mathbb{P}(M > x)$ with a given accuracy for each given $x \geq 0$. In this regard we consider the difference

$$\mathbb{P}(M > x) - \mathbb{P}(M(t) > x) = \mathbb{P}(M > x,\, M(t) \leq x) =: \gamma(x, t) \geq 0.$$

There are some possible probabilistic distances based on $\gamma(x, t)$, but we will focus on Kolmogorov-Smirnov (uniform) distance,

$$D(t) := \sup_{x > 0} \gamma(x, t),$$

which measures the maximum distance between distributions. (Another popular integral distance, or L_1-distance, measures the total distance between the distributions [11].)

The main contribution of this paper is that we establish convergence rate to stationarity in the queueing system fed by a general Gaussian input, and it is an extension of the corresponding result from [11] proved for the system with fB input. We mainly follow the approach developed in [3, 11], and it allows us to focus only on the analysis of the differences in the corresponding proofs.

2   Large Deviation Background

First we consider an important particular case $A = B_H$, that is the fB input
with covariance function

$$\Gamma(s,t) := \mathbb{E}\Big[B_H(s)B_H(t)\Big] = \frac{1}{2}\left(|t|^{2H} + |s|^{2H} - |t-s|^{2H}\right).$$

We recall the Large Deviation Principle (LDP) framework which is used throughout the paper. Define function space Ω of the trajectories of process B_H,

$$\Omega = \left\{\omega \in C(\mathbb{R}),\ \omega(0) = 0,\ \lim_{|t|\to\infty}\frac{\omega(t)}{1+|t|} = 0\right\},$$

with the norm

$$\|\omega\|_\Omega := \sup_{t\in\mathbb{R}}\frac{|\omega(t)|}{1+|t|}.$$

The *reproducing kernel Hilbert space* $\mathcal{R}$ (RKHS) associated with the distribution
of B_H is the closure of the following set of functions

$$f := \sum_{k=1}^{n} f_k\,\Gamma(t_k,\cdot),\ f_k \in \mathbb{R},\ n \in \mathbb{N}.$$

The inner product of $f,\ g \in \mathcal{R}$ is defined as

$$\langle f, g\rangle_\mathcal{R} = \sum_{i=1}^{n}\sum_{j=1}^{m} f_i\,g_j\Gamma(t_i,t_j).$$

On the space Ω, define the *rate function*

$$I(f) = \begin{cases} \frac{1}{2}\|f\|_\mathcal{R}^2, & f \in \mathcal{R}, \\ \infty, & \text{otherwise,} \end{cases} \tag{8}$$

where $\|f\|_\mathcal{R} := \sqrt{\langle f, f\rangle_\mathcal{R}}$. Function $I(f)$ can be interpreted as a measure for the
likelihood of a path, and the path $f^* := \arg\inf I(f)$ is called the *most likely path*.
In what follows, we need the following generalized version of Schilder's theorem
(LDP for fBm) [14]:

$$\limsup_{n\to\infty}\frac{1}{n}\ln\mathbb{P}\left(\frac{A}{\sqrt{n}} \in F\right) \leq -\inf_{f\in F} I(f),$$

for any closed set $F \in \Omega$, and

$$\liminf_{n\to\infty}\frac{1}{n}\ln\mathbb{P}\left(\frac{A}{\sqrt{n}} \in G\right) \geq -\inf_{f\in G} I(f),$$

for any open set $G \in \Omega$.

A set F is called *good* (with respect to the rate function I) if

$$\inf_{f \in F^\circ} I(f) = \inf_{f \in \overline{F}} I(f),$$

where F° and $\overline{F}$ are the interior and the closure of the set F, respectively. For any good set F, generalized version of Schilder's theorem states informally that

$$\mathbb{P}\left(\frac{A}{\sqrt{n}} \in F\right) \approx \exp\left(-n \inf_{f \in F} I(f)\right),$$

i. e., the decay rate is dominated by the path with minimal $\mathcal{R}$-norm (the most likely path).

2.1   LDP with Appropriate Scaling

Consider Gaussian process A with variance (1) and, for each $\alpha > 0$, define the following scaled process $A^{(\alpha)} = \{A_t^{(\alpha)}, \, t \in \mathbb{R}\}$, where

$$A_t^{(\alpha)} := \frac{A_{\alpha t}}{\sigma(\alpha)}.$$

Denote

$$v(\alpha) = \frac{\alpha^2}{\sigma^2(\alpha)}.$$

It has been shown in [3] that, under the (additional) assumption that the following limit exists

$$\lim_{t \to 0} \sigma^2(t)|\log|t||^{1+\varepsilon} < \infty \text{ for some } \varepsilon > 0, \tag{9}$$

the pair

$$\left(\left(\frac{A^{(\alpha)}}{\sqrt{v(\alpha)}}, v(\alpha)\right) : \alpha \geq 1\right)$$

satisfies a LDP with good rate function, i. e.,

$$\limsup_{\alpha \to \infty} \frac{1}{v(\alpha)} \ln \mathbb{P}\left(\frac{A^{(\alpha)}}{\sqrt{v(\alpha)}} \in F\right) \leq -\inf_{f \in F} I(f), \tag{10}$$

for any closed set $F \in \Omega$, and

$$\liminf_{\alpha \to \infty} \frac{1}{v(\alpha)} \ln \mathbb{P}\left(\frac{A^{(\alpha)}}{\sqrt{v(\alpha)}} \in G\right) \geq -\inf_{f \in G} I(f) \tag{11}$$

for any open set $G \in \Omega$. Moreover the rate function I is defined by the RKHS associated with the corresponding fBm (see [4] p. 124, eq. (2.9) and [3], p. 867, Corollary 3), i. e., satisfies definition (8).

Remark. Under extra technical conditions, the LDP (10), (11) has been firstly established by Kozachenko et al. [4], and then Dieker [3] has weakened these conditions up to (9).

3   Related Asymptotic Results

Busy period asymptotics. A LPD approach described above allows to obtain asymptotic results for various stationary measures, in particular, for an ongoing stationary busy period

$$K := \inf\{t \geq 0 : Q(t) = 0\} - \sup\{t \leq 0 : Q(t) = 0\}.$$

Dieker [3] has shown that for the input process with variance (9),

$$\lim_{t \to \infty} \frac{\sigma^2(t)}{t^2} \log \mathbb{P}(K > t) = -\inf_{f \in \mathscr{B}} I(f) := -\nu, \tag{12}$$

where the set of functions

$$\mathscr{B} := \{f \in \mathcal{R} : f(r) \geq r, \forall r \in [0,1]\},$$

and the rate function I is determined by (8). Because the kernel $\mathcal{R}$ is defined by the distribution of the corresponding fBm, then the constant ν in (12) depends on the index variation $2H$ only. Relation (12) has been firstly established by Norros [17] for $A = B_H$, and then it was generalized by Kozachenko et al. [4] and Dieker [3]. Moreover, it is known that the constant $\nu \in [\frac{1}{2}, c_H^2/2]$, where

$$c_H := \left[H(2H-1)(2-2H) \, \mathsf{B}(H-1/2, 2-2H) \right]^{-1/2}, \tag{13}$$

and B is the Beta function [17]. A characterization of the most likely path in the set $\mathscr{B}$ has been found in [12], and, because explicit expression for ν is not available, the numerical methods to calculate ν have been proposed.

Convergence rate asymptotics. There is a close connection between the busy period asymptotics and the asymptotics of the convergence rate. In particular, the following logarithmic asymptotics for fBm input holds [11]:

$$\lim_{t \to \infty} \frac{1}{t^{2-2H}} \log D(t) = -\inf_{f \in \mathscr{D}} I(f) = -\nu,$$

i. e., the decay rate of $D(t)$ is exactly the same, in the logarithmic sense, as the decay rate of the probability $\mathbb{P}(K > t)$ in (12).

4   Main Asymptotic Result

A LDP with an appropriate scaling allows to extend previous result to more general Gaussian input. Namely, we prove the following statement.

Theorem 1. *For a general Gaussian input with regularly varying variance* (1),

$$\lim_{t \to \infty} \frac{1}{v(t)} \log D(t) = -\nu. \tag{14}$$

Proof. The proof is mainly based on arguments from [11] and [3]. For any $x > 0$, $t \geq 0$ we have

$$\gamma(x,t) = \mathbb{P}(M > x, M(t) \leq x)$$

$$= \mathbb{P}\left(\forall r \in [0,t] : A_r \leq x + r; \; \exists s > t : A_s > x + s\right)$$

$$= \mathbb{P}\left(\forall r \in [0,1] : \frac{A_r^{(t)}}{\sqrt{v(t)}} \leq \frac{x}{t} + r; \; \exists s > 1 : \frac{A_s^{(t)}}{\sqrt{v(t)}} > \frac{x}{t} + s\right).$$

Observe that,

$$\liminf_{t \to \infty} \frac{1}{v(t)} \log D(t) = \liminf_{t \to \infty} \frac{1}{v(t)} \log \sup_{x > 0} \gamma(x,t)$$

$$= \liminf_{t \to \infty} \frac{1}{v(t)} \log \sup_{x > 0} \mathbb{P}\left(\forall r \in [0,1] : \frac{A_r^{(t)}}{\sqrt{v(t)}} \leq x + r; \right.$$

$$\left. \exists s > 1 : \frac{A_s^{(t)}}{\sqrt{v(t)}} > x + s\right). \quad (15)$$

For each $x \geq 0$, denote

$$\mathbb{P}_t(x) = \mathbb{P}\left(\forall r \in [0,1] : \frac{A_r^{(t)}}{\sqrt{v(t)}} \leq x + r; \; \exists s > 1 : \frac{A_s^{(t)}}{\sqrt{v(t)}} > x + s\right).$$

Then it follows that for arbitrary $\varepsilon > 0$,

$$\liminf_{t \to \infty} \frac{1}{v(t)} \log D(t) = \liminf_{t \to \infty} \frac{1}{v(t)} \log \sup_{x > 0} \mathbb{P}_t(x)$$

$$\geq \liminf_{t \to \infty} \frac{1}{v(t)} \log \mathbb{P}_t(\varepsilon). \quad (16)$$

For each $\varepsilon \geq 0$ define the set

$$\mathscr{A}_\varepsilon := \{f \in \mathcal{R} \mid \forall r \in [0,1] : f(r) \leq \varepsilon + r; \; \exists s > 1 : f(s) > \varepsilon + s\}.$$

It is easy to see that $\mathscr{A}_\varepsilon$ is an open subset of Ω. It then follows from LDP (11) that

$$\liminf_{t \to \infty} \frac{1}{v(t)} \log \mathbb{P}_t(\varepsilon) \geq - \inf_{f \in \mathscr{A}_\varepsilon} I(f).$$

Moreover, according to proposition 3.3 of [11],

$$\inf_{f \in \mathscr{A}_0} I(f) = \inf_{f \in \mathscr{B}} I(f) = \nu. \quad (17)$$

Now letting $\varepsilon \downarrow 0$ in (16) and applying (17) we obtain

$$\lim_{t \to \infty} \frac{1}{v(t)} \log D(t) \geq -\nu,$$

so the lower bound in (14) is established. Because all corresponding properties of the fBm path space remain in force in our setting, then, to establish the upper bound in (14), we again can apply arguments from [11]. More exactly, we have as in (15) that

$$\limsup_{t\to\infty} \frac{1}{v(t)} D(t) = \limsup_{t\to\infty} \frac{1}{v(t)} \log \sup_{x>0} \mathbb{P}_t(x).$$

For a fixed $\varepsilon > 0$, we now take arbitrary $u \in \mathbb{R}$ which is multiply $\varepsilon > 0$. It gives

$$\sup_{x>0} \mathbb{P}_t(x) \le \sum_{k=1}^{u/\varepsilon} \mathbb{P}_t^\varepsilon(k) + \sup_{x>u} \mathbb{P}_t(x), \tag{18}$$

where $\mathbb{P}_t^\varepsilon(k) := \sup_{(k-1)\varepsilon \le x \le k\varepsilon} \mathbb{P}_t(x)$. Consider the second term on the right in (18). It is easy to see that (for arbitrary x)

$$\sup_{x>u} \mathbb{P}_t(x) \le \sup_{x>u} \mathbb{P}\left(\exists s > 1 : \frac{A_s^{(t)}}{\sqrt{v(t)}} > x+s\right) = \mathbb{P}\left(\exists s > 1 : \frac{A_s^{(t)}}{\sqrt{v(t)}} > u+s\right).$$

If we take now $u \ge H^{-1} - 1$, then by the proof of Lemma 3.5 in [11],

$$\limsup_{t\to\infty} \frac{1}{v(t)} \log \mathbb{P}\left(\exists s > 1 : \frac{A_s^{(t)}}{\sqrt{v(t)}} > u+s\right) \le -\nu. \tag{19}$$

The kth summand in the first term on the right in (18) is upper bounded as

$$\mathbb{P}_t^\varepsilon(k) \le \mathbb{P}\left(\forall r \in [0,1] : \frac{A_r^{(t)}}{\sqrt{v(t)}} \le k\varepsilon + r; \exists s \ge 1 : \frac{A_s^{(t)}}{\sqrt{v(t)}} \ge (k-1)\varepsilon + s\right).$$

Define the set

$$\mathscr{A}_{x,\varepsilon} := \{f \in \mathcal{R} \,|\, \forall r \in [0,1] : f(r) \le x+r; \exists s \ge 1 : f(s) \ge x - \varepsilon + s\}.$$

It is not difficult to show that $\mathscr{A}_{x,\varepsilon}$ is closed in Ω. Hence, by the LDP (10), we obtain

$$\limsup_{t\to\infty} \frac{1}{v(t)} \log \mathbb{P}_t^\varepsilon(k) \le -\inf_{f \in \mathscr{A}_{k\varepsilon,\varepsilon}} I(f). \tag{20}$$

It then follows from (19)-(20)

$$\limsup_{t\to\infty} \log \mathbb{P}_t(x) \le \max\left\{\max_{x=\varepsilon,2\varepsilon,\dots,u} \left(-\inf_{f \in \mathscr{A}_{x,\varepsilon}} I(f)\right), -\nu\right\}. \tag{21}$$

Now, as in [11] (pp. 1395–1396), we obtain

$$-\inf_{f \in \mathscr{A}_{x,\varepsilon}} I(f) \le -\inf_{f \in \mathscr{A}_{0,\varepsilon}} I(f),$$

and thus, the the right-hand side of (21) is upper bounded by

$$\max\left\{-\inf_{f\in\mathscr{A}_{0,\varepsilon}} I(f), -\nu\right\}.$$

Letting $\varepsilon \downarrow 0$ gives the following upper bound in (14),

$$\limsup_{t\to\infty} \frac{1}{v(t)} D(t) \leq \max\{-\inf_{f\in\mathscr{A}_{0,0}} I(f), -\nu\}.$$

Finally, using standard continuity arguments (see, for instance, Section 4 in [17] or Appendix in [13]), one can show that

$$\inf_{f\in\mathscr{A}_{0,0}} I(f) = \inf_{f\in\mathscr{A}_0} I(f) = \nu,$$

and it completes the proof.

5   Estimation of the Simulation Horizon

In this section we describe in brief estimation of the overflow probability $\mathbb{P}(M > x)$ via simulation, i.e., the estimation of the probability that stationary workload exceeds some threshold x. To this end, we must generate a number of the trajectories of the Gaussian input with length T, and the main question is: how large T must be? Actually, in this case, instead of $\mathbb{P}(M > x)$, the probability $\mathbb{P}(M(T) > x)$ is estimated. Define

$$\tau_x = \inf\{t \geq 0 : A_t - t \geq x\},$$

then evidently, $\mathbb{P}(M(T) > x) = \mathbb{P}(\tau_x \leq T)$, implying

$$\mathbb{P}(M > x) = \mathbb{P}(\tau_x < \infty) = \mathbb{P}(\tau_x \leq T) + \mathbb{P}(T < \tau_x < \infty).$$

The last expression shows that, in order to approximate $\mathbb{P}(M > x)$ by $\mathbb{P}(M(T) > x)$ with given accuracy $\epsilon > 0$, we must take T so large that

$$\frac{\mathbb{P}(T < \tau_x < \infty)}{\mathbb{P}(\tau_x < \infty)} < \epsilon, \tag{22}$$

see Chapter 8 of [14] for more details. Further, note that

$$\mathbb{P}(T < \tau_x < \infty) = \gamma(x, T) \leq D(T).$$

Moreover, according to Theorem 1, for all T sufficiently large,

$$D(T) \approx \exp(-\nu v(T)).$$

Now consider the denominator of (22):

$$\mathbb{P}(\tau_x < \infty) = \mathbb{P}(M > x) = \mathbb{P}\left(\sup_{s \geq 0}(A_s - s) > x\right)$$

$$\geq \sup_{s \geq 0} \mathbb{P}(A_s > x + s)$$

$$= \sup_{s \geq 0} \Psi\left(\frac{x+s}{\sigma(s)}\right) := G(x), \qquad (23)$$

where Ψ denotes the tail distribution of standard normal variable,

$$\Psi(t) = \frac{1}{\sqrt{2\pi}} \int_t^\infty e^{-y^2/2} dy.$$

The function $G(x)$ is not explicitly available for a general Gaussian input, while it is known for fBm [16]:

$$G(x) = \Psi\left(\frac{x + \bar{s}}{\sigma(\bar{s})}\right), \quad \text{where } \bar{s} = \frac{xH}{1-H}.$$

Thus, in general case $G(x)$ can be calculated numerically. In this case the left hand side of (22) is approximated by

$$\frac{\exp(-\nu\, v(T))}{G(x)}.$$

Now the lower bound of the required sample length can be found as the minimal T satisfying the inequality

$$v(T) \geq -\frac{1}{\nu} \log(\epsilon\, G(x)). \qquad (24)$$

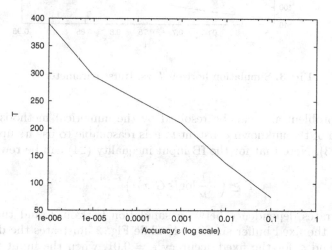

Fig. 1. Simulation horizon T vs. accuracy ϵ

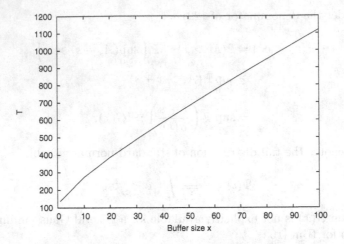

Fig. 2. Simulation horizon T vs. buffer size x

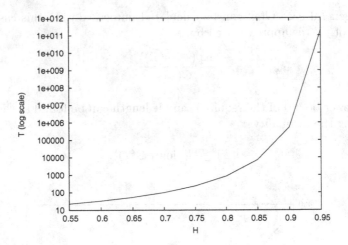

Fig. 3. Simulation horizon T vs. Hurst parameter H

The latter problem also can be resolved by the numerical methods, in which case, instead of the unknown constant ν, it is reasonable to use its upper bound $c_H^2/2$, see (13). Note that for the fB input inequality (24) can be rewritten as

$$T \geq \left(-\frac{1}{\nu} \log(\varepsilon\, G(x)) \right)^{\frac{1}{2-2H}}. \tag{25}$$

Fig. 1 illustrates dependence between simulation horizon T and the accuracy (error) ϵ for the fixed buffer size $x = 1$, while Fig. 2 illustrates the dependence between T and x for the fixed accuracy $\epsilon = 0.01$, when the input is the sum of two independent fB inputs with parameters $H_1 = 0.7$, $H_2 = 0.6$. Finally,

Fig. 3 shows dependence between T and parameter H, corresponding to the lower bound (25), for a single fB input and $x = 1$.

6 Conclusion

We consider the rate of convergence to stationarity of the fluid queueing system with a constant service rate and with Gaussian input with stationary increments. The variance of the input is regularly varying with index $2H$, $0 < H < 1$. For this system, we prove that the convergence rate is similar to that has been earlier obtained in [11] for the fluid system with the corresponding fB input. We also discuss estimation of the overflow probability by simulation and present some numerical examples.

References

1. Addie, R., Mannersalo, P., Norros, I.: Most probable paths and performance formulae for buffers with Gaussian input traffic. European Transactions in Telecommunications. **13**, 183–196 (2002)
2. Anick, D., Mitra, D., Sondhi, M.M.: Stochastic theory of a data handling system with multiple resources. Bell System Technical Journal. **61**, 1871–1894 (1982)
3. Dieker, A.B.: Conditional limit theorems for queues with Gaussian input: a weak convergence approach. Stochastic Processes and their Applications. **115**(5), 849–873 (2005)
4. Kozachenko, Y., Vasylyk, O., Sottinen, T.: Path space large deviations of a large buffer with Gaussian input traffic. Queueing Syst. **42**, 113–129 (2002)
5. Duffield, N., O'Connell, N.: Large deviations and overflow probabilities for the general single server queue, with applications. Proceedings of the Cambridge Philosophical Society. **118**, 363–374 (1995)
6. Debicki, K.: A note on LDP for supremum of Gaussian processes over infinite horizon. Stat. Probab. Lett. **44**, 211–220 (1999)
7. Hüsler, J., Piterbarg, V.: Extremes of a certain class of Gaussian processes. Stochastic Process. Appl. **83**, 257–271 (1999)
8. Kilpi, J., Norros, I.: Testing the Gaussian approximation of aggregate traffic. In: Proceedings of the 2nd Internet Measurement Workshop, pp. 49–61 (2002)
9. Kulkarni, V., Rolski, T.: Fluid model driven by an Ornstein-Uhlenbeck process. Probability in the Engineering and Informational Sciences. **8**, 403–417 (1994)
10. Leland, W.E., Taqqu, M.S., Willinger, W., Wilson, D.V.: On the self-similar nature of ethernet traffic (extended version). IEEE/ACM Transactions of Networking. **2**(1), 1–15 (1994)
11. Mandjes, M., Norros, I., Glynn, P.: On convergence to stationarity of fractional Brownian storage. Ann. Appl. Probab. **19**(4), 1385–1403 (2009)
12. Mandjes, M., Mannersalo, P., Norros, I., van Uitert, M.: Large deviations of infinite intersections of events in Gaussian processes. Stochastic Process. Appl. **116**, 1269–1293 (2006)
13. Mandjes, M., van Uitert, M.: Sample-path large deviations for tandem and priority queues with Gaussian inputs. Ann. Appl. Probab. **15**, 1193–1226 (2005)
14. Mandjes, M.: Large Deviations for Gaussian Queues: Modelling Comminication Networks. Wiley, Chichester (2007)

15. van de Meent, R., Mandjes, M., Pras, A.: Gaussian traffic everywhere? In: Proceedings of IEEE International Conference on Communications (ICC), vol. 2, pp. 573–578 (2006)
16. Norros, I.: A storage model with self-similar input. Queueing Syst. **16**, 387–396 (1994)
17. Norros, I.: Busy periods for fractional Brownian storage: a large deviation approach. Adv. in Perf. Anal. **2**(1), 1–19 (1999)
18. Reich, E.: On the integrodifferential equation of Takacs I. Ann. Math. Stat. **29**, 563–570 (1958)
19. Taqqu, M.S., Willinger, W., Sherman, R.: Proof of a fundamental result in self-similar traffic modeling. Computer communication review. **27**, 5–23 (1997)
20. Willinger, W., Taqqu, M.S., Leland, W.E., Wilson, D.: Self-similarity in high-speed packet traffic: analysis and modeling of Ethernet traffic measurements. Statistical Sciences. **10**(1), 67–85 (1995)

Model-Based Quantitative Security Analysis of Mobile Offloading Systems Under Timing Attacks

Tianhui Meng[✉], Qiushi Wang, and Katinka Wolter

Department of Mathematics and Computer Science,
Freie Universität Berlin, Takustr. 9, 14195 Berlin, Germany
{tianhui.meng,qiushi.wang,katinka.wolter}@fu-berlin.de

Abstract. Mobile offloading systems have been proposed to migrate complex computations from mobile devices to powerful servers. While this may be beneficial from the performance and energy perspective, it certainly exhibits new challenges in terms of security due to increased data transmission over networks with potentially unknown threats. Among possible security issues are timing attacks which are not prevented by traditional cryptographic security. Metrics on which offloading decisions are based must include security aspects in addition to performance and energy-efficiency. This paper aims at quantifying the security attributes of mobile offloading systems. The offloading system is modeled as a stochastic process. The security quantification analysis is carried out for steady-state behaviour as to optimise a combined security and cost trade-off measure.

Keywords: Mobile offloading · Security attributes · Quantitative analysis · Semi-Markov process

1 Introduction

Cloud computing has become widely accepted as computing infrastructure of the next generation, as it offers advantages by allowing users to exploit platforms and software provided by cloud providers (e.g., Google, Amazon and IBM) from anywhere on demand at low price [1]. At the same time, mobile devices are progressively becoming an important constituent part of everyday life as very convenient communication and business tools with a wide variety of software covering all aspects of life. Mobile devices allow to make transactions in almost every possible situation in life, even while walking on the street. The concept of computation offloading has been proposed with the objective to migrate large computations and complex processing from mobile devices with energy limitations to resourceful servers in the cloud. This avoids a long application execution time on mobile devices which results in large power consumption.

Over the last years, research on computation offloading focussed on how to offload and what to offload from mobile devices to cloud servers in order to

© Springer International Publishing Switzerland 2015
M. Gribaudo et al. (Eds.): ASMTA 2015, LNCS 9081, pp. 143–157, 2015.
DOI: 10.1007/978-3-319-18579-8_11

reduce the execution time and power consumption of computation tasks [2]. Several offloading infrastructures have been developed for offloading at varying granularity, among which the MAUI offloading system, presented in 2010, not only achieves significant reduction in energy consumption for some jobs on mobile devices, but also improves the performance of mobile applications (i.e., refresh rate of a game can increase from 6 to 13 frames per second) [3]. In addition, instead of offloading the full code, MAUI partitions the application code at runtime to maximize energy savings. However, several challenges still exist in the following three aspects of mobile offloading systems:

Time and energy consumption in data transition

Data transmission over wireless or cellular networks is of highly unpredictable quality. Wu [4] proposed metrics to express the energy response time tradeoff, the Energy-Response time Weighted Sum (ERWS) and Energy-Response time Product (ERP) for mobile offloading systems which can be optimised using different offloading policies.

Lossy network

Low bandwidth or long delays are a possible factor incurring network unreliability. Consequently, when migrating the computation to the cloud server, the execution of the offloading task may suffer from long delays or even failures by the unreliable network. Limited battery capacity of the mobile device prohibits unpredictable waiting times, which may also be caused by a long recovery process. A dynamic scheme to determine whether and when to launch the local re-execution, instead of always waiting for network recovery to offload [5] may help to deal with this problem.

Security and data confidentiality

Along with the benefits of high performance, the offloading system witnesses potential security threats including compromised data due to the increased number of parties, devices and applications involved, that leads to an increase in the number of points of access. Security threats have become an obstacle in the rapid expansion of the mobile cloud computing paradigm. Significant efforts have been devoted in research organisations and academia to build secure mobile cloud computing environments and infrastructures [6]. However, work on modelling and quantifying the security attributes of mobile offloading system is rare.

Quantitative analyses of system dependability and reliability have received great attention for several decades. In 1993, Littelwood [7] first introduced the idea to evaluate the system security attributes using analytical methods of system reliability. Then, Nicol et al. [8] surveyed the model-based techniques for evaluating system dependability, and summarized how they can be extended to evaluate system security. Previous work on the security of computing and information systems has been mostly assessed from a level point of view. A system is assigned a given security level with respect to the presence or absence of certain functional characteristics and the use of certain development techniques. In 2013, Zhang [9] proposed an approach to evaluate the network security situation objectively using Network Security Index System (NSIS). Only a few studies have considered the quantitative evaluation of security. The authors in [10] make an effort to examine the security vulnerabilities of operating systems of routers within the cloud carrier by assessing the risk based on the National Vulnerability Database (NVD) and gives a quantifiable security metrics for cloud carrier, which is very useful in the Service Level Agreement (SLA) negotiation between a cloud consumer and a cloud provider.

In this paper, we propose a state transition model of a general mobile offloading system under the specific threat of timing attacks. Our model is aimed to deal with a general offloading system with a master secret stored on the server side, where the timing attacker can get normal offloading service. In a timing attack the attacker deduces information about a secret key from runtime measurements of successive requests. This process can be interrupted by frequently changing the key [11]. From the security quantification point of view, since the sojourn time distribution function in different system states may not always be exponential, the underlying stochastic model needs to be formulated as a Semi-Markov Process (SMP). Computing the combined system security and cost trade-off metric, we investigate the cost for a given security requirement. Our results will give security metrics on which offloading decisions are based. To enhance system security, the sensitivity of the influencing factor in the quantitative analysis is also discussed.

The remainder of this paper is structured as follows. In Section 2, we develop a Semi-Markov model for a general offloading system under the threat of timing attack. The steady-state probabilities leading to the computation of steady-state security measures is addressed in Section 3. Section 4 shows numeral results of the analysis performed on the model for a sample. And finally, the paper is concluded and future work are presented in Section 5.

2   Security Analysis Based on SMP Model

A mobile offloading system is a solution to enhance the capabilities of the mobile system by migrating computation to more resourceful computers (i.e., servers). To quantitatively analyse the security attributes of a system under the threat of timing attacks, we have to incorporate the actions of an attacker who is trying to capture sensitive information in conjunction with the protective actions taken by the offloading system. Therefore, we have to develop a composite security model that takes into account the behaviour of both actors. Semi-Markov Processes (SMPs) are generalizations of Markov chains where the sojourn times in the states need not be exponentially distributed [12].

The state transition model represents the system behaviour for a specific attack and given system configuration that depends on the actual security requirements. In our scenario, the system is assumed to be vulnerable to timing attacks in which the attacker in the worst case will eventually decrypt the system key. We assume that the server is configured as to renew its key regularly to prevent or handle these attacks.

2.1 Behaviour of Attacker and System

Timing attacks gain information from the server response time and rather than brute force attacks or theoretical weaknesses in the algorithms they are a real threat to mobile offloading systems. However this threat is not covered by traditional notions of cryptographic security [13]. It was commonly believed that timing attacks can be directed only towards smart cards or affect inter-process locally, but more recent research reveals that remote timing attacks are also possible and should be taken into consideration [14][15]. Mobile offloading requires access to resourceful servers for short durations through wireless networks. These servers may use virtualization techniques to provide services so that they can isolate and protect different programs and their data. However, the author in [16] shows that using a cache timing attack, an attacker can bypass the isolated environment provided by virtualisation characteristics, where sensitive code is executed in isolation from untrustworthy applications. It is worth mentioning that a timing attack also poses a threat to other types of systems. Timing attacks can be detrimental in the mix-zone construction and usage model over road networks [17].

In the offloading systems we consider, a server master key is used for the encryption and decryption operations of user data. In order to improve security, the server should regularly or irregularly change the master key. The system has to process all user-files with both the new and the old master key. In this process, the system does not accept any other user commands. When user data is very large, this process will take long. Therefore, it is reasonable to recommend a minimum time for the master key replacement cycle, and select a suitable time, which is when there is a low amount of user access(e.g.. at night).

In timing attacks to our offloading system, an attacker will continue to send requests to the server and the obtained service will be properly performed by the server. In addition the attacker records each response time for a certain service and tries to find clues to the master secret of the server by comparing time differences from several request queues. If the attacker successfully breaks the secret information from the timing results, he can read and even modify other users information without authorisation.

2.2 The Model

Fig. 1 depicts the state transition model we propose for describing the dynamic behaviour of a generic offloading system. This system is under the specific threat of timing attacks conducted by random attackers. We describe the events that trigger transitions among states in terms of probabilities and cumulative distribution functions, which will be shown later.

The states and parameters of the SMP model are summarized here:

- I Initial state of the offloading system after star up
- T Timing attack happening state
- A Attack state after the attacker get the secret of the system

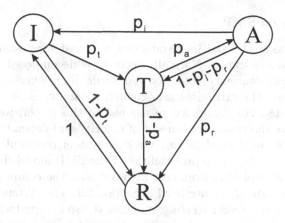

Fig. 1. State transition diagram for a generic offloading system

- R Rekeying state
- p_t probability that an attacker begin to conduct a timing attack to the system
- p_a probability of attack system confidentiality after a successful timing attack
- p_i probability that the system return initial state by manual intervention
- p_r probability that the attack is terminated due to rekeying operation

After initialisation, the system is in the good state I. The sojourn time in state I is the life time of the system before an attacker starts a timing attack or the system renews its key. We assume there is only one attacker in the system at one time. If an attack happens, the system is brought to state T, in which the timing attack takes place and the attacker deciphers the encryption key by making time observations. So while the system is in state T, the attacker is not yet able to access confidential information. We assume that it takes a certain time to perform the timing attack after which the attacker will know the encryption key and the system moves to the compromised state A. Changing the encryption key can prevent or interrupt a timing attack. During rekeying the system is in state R. The challenge is to find an optimal value for the rekey interval. The rekeying should certainly happen before or soon after the system enters the compromised state. Rekeying will bring the system back to the initial state I.

If the attacker succeeds to determine the encryption key through time measurements confidential data will be disclosed which is assumed to incur a high cost. This can only happen if the system is in the compromised state A and we call the incident of entering the compromised state a security failure. One possibility is that one attacker stops himself and another attacker comes for a new timing attack. So the system is brought from compromised state A to another timing attack state T. The attack can also be stopped by manual intervention, i.e. triggering the rekey operation. This can happen either in the attack state T or in the compromised state A, both transitioning the system to the rekey state R from which it will return to the initial state.

2.3   Measures on SMP

After defining the model and its parameters, we must now establish the measures we want to investigate. Normally security is decomposed into three different aspects: confidentiality, integrity and availability, whereas dependability is decomposed into the attributes: availability, reliability, safety, integrity and maintainability [18]. For simplicity, in the rest of this paper, we use the term security to denote the combined concept of security and dependability.

The measures are defined in this work as system cost and confidentiality that are functions of the state probabilities of the SMP model. In our scenario, the offloading system suffers from cost in two states, the compromised state A and the rekeying state R, as discussed in Section 2.1. The system loses sensitive information in the compromised state, and cost is also incurred when the system deploys a rekeying process regularly. The steady-state probabilities π_i may be interpreted as the proportion of time that the SMP spends in the state i. In our model, the rekeying cost and the data disclosed cost are both interpreted as the proportion of system life time, that is, the steady-state probability of the SMP. We define two weights c and its complement $1 - c$ for the two kinds of cost. We use normalization weights for simplicity. The system cost is defined as:

$$Cost = c\pi_A + (1 - c)\pi_R . \tag{1}$$

where $\pi_i, i \in \{A, R\}$ denotes the steady-state probability that the SMP is in state i. $0 \leq c \leq 1$ is the weighting parameter used to share relative importance between the loss of sensitive information and the effort needed to rekey regularly. Similarly, if a timing attack to the offloading system is successful, the attacker obtains the master key and can browse unauthorised files thereafter. The entered states denote the loss of confidentiality. Therefore, the steady-state confidentiality measure can then be computed as

$$Confid = 1 - \pi_A . \tag{2}$$

In order to investigate how system security will interact with the cost, we also define a trade-off metric. An objective function formed from the division of the security attribute confidentiality and system cost is created to demonstrate the relationship between the cost the system has to pay and the corresponding security system gain. The trade-off metric shows the how much security per cost you can obtain. As a system designer, one may look forward to maintaining the confidentiality of sensitive informations with lower system cost, as for the trade-off measure, the larger the better.

$$Trade = \frac{Confid}{Cost} . \tag{3}$$

In the next two Sections, we will evaluate these measures by computing the steady-state probability of the SMP model and synthesize the effect of parameter changing by sensitivity analysis.

3   Semi-Markov Process Analysis

In this section, we derive and evaluate the security attributes using methods for quantitative assessment of dependability, known as the dependability attributes, e.g. reliability, availability, and safety which have been well established quantitatively.

For the offloading system, we have described the system's dynamic behaviour by a SMP model with the states $\{I, T, A, R\}$ and the transition between these states. A system response to a security attack is fairly automated and could be quite similar to how it may respond to accidental faults. Let $\{X(t) : t \geq 0\}$ be the underlying stochastic process with a discrete state space $X_s = \{I, T, A, R\}$. To obtain a complete description of this SMP model, two sets of parameters must be known: the mean sojourn time h_i in each state and the transition probabilities p_{ij} between different states, where $i, j \in X_s$, which we have depicted in the previous Section.

3.1   DTMC Steady-State Probability Computations

It was explained earlier in order to carry out the security quantification analysis, we need to analyse the SMP model of the system that was described by its state transition diagram. The steady-state probabilities $\{\pi_i, i \in X_s\}$ of the SMP states are computed in terms of the embedded DTMC steady-state probabilities v_i and the mean sojourn times h_i[19]:

$$\pi_i = \frac{v_i h_i}{\sum_j v_j h_j} \quad i, j \in X_s. \tag{4}$$

Assuming the existence of the steady-state in the underlying DTMC, it can be computed as

$$\vec{v} = \vec{v} \cdot \mathbf{P} \quad i \in X_s. \tag{5}$$

where $\vec{v} = [v_I, v_T, v_A, v_R]$ and $\mathbf{P}$ is the DTMC transition probability matrix which can be written as:

$$
\mathbf{P} = \begin{array}{c} \\ I \\ T \\ A \\ R \end{array}
\begin{array}{cccc}
I & T & A & R
\end{array}
\left(
\begin{array}{cccc}
0 & p_t & 0 & 1 - p_t \\
0 & 0 & p_a & 1 - p_a \\
p_i & 1 - p_i - p_r & 0 & p_r \\
1 & 0 & 0 & 0
\end{array}
\right) \tag{6}
$$

In addition, we have the total probability relationship:

$$\sum_i v_i = 1 \quad i \in X_s. \tag{7}$$

The transition probability matrix $\mathbf{P}$ describes the DTMC state transition probabilities between the DTMC states as shown in Fig. 1. The first step towards evaluating security attributes is to find the steady-state probability vector $\vec{v}$ of the DTMC states by solving Eqs. 5 and 7. We can get solutions:

$$v_I = \frac{p_i p_a + 1 - p_a + p_a p_r}{\phi}, \tag{8}$$

$$v_T = \frac{p_t}{\phi}, \; v_A = \frac{p_t p_a}{\phi}, \; v_R = v_I - \frac{p_i p_t p_a}{\phi}$$

For the sake of brevity, we assume: $\phi = 2 + 2p_i p_a + p_t + p_t p_a - 2p_a + p_a p_r - p_i p_t p_a$.

Note that the analysis carried out in this paper depends only on the mean sojourn time and is independent of the actual sojourn time distributions for the SMP states. If we were to carry out a transient analysis of the SMP, this would no longer be true. In the next subsection, the DTMC steady-state probabilities are used to compute the SMP steady-state probabilities.

3.2 Semi-Markov Model Analysis

The mean sojourn time h_i in a particular state $i \in X_s$ is the other quantity that is needed to compute the SMP steady-state probabilities. It is determined by the random time that a process spends in a particular state as discussed in Section 3. In the computer security domain, there is a wide variety of attackers ranging from amateur hackers to inimical intelligence agencies possessing a wide spectrum of expertise and resources.

The parameters h_T, h_A, p_t, p_a depend on the attackers' behavior which we assumed as random processes. The analysis in this paper only takes into account the mean value of these processes. More complex study will consider a quantitative analysis of attacker behavior based on empirical data. However, this paper limits itself to dealing with an SMP model only. The attacker model is planned to be covered separately in a future paper. We put the h_i again here:

- h_I the mean time the system spends before an attacker conducts a timing attack or rekey itself
- h_T the mean time before the attacker break the master secret of the server by timing attack
- h_A the mean time the system is losing information
- h_R the mean time for rekeying process

Clearly, for the model to be accurate, it is important to estimate accurately the model parameters. Some parameters we will get from experiments. The measurements we are in process of taking are based on a offloading server under timing attacks. We have built a timing attack demonstrator and measure the mean time for a successful attack which will be used as h_I. Some parameters, e.g. probability that an attacker begin to conduct a timing attack and attack system confidentiality after a successful timing attack will be assumed as an attacker. Other parameters used in our system can be tune by the system administrator, like the rekey probability p_r and the mean sojourn time in initial state h_I.

In this work, however, our focus is primarily on developing a quantitative analysis methodology for the security attributes of an offloading system. So, in the absence of exact values of model parameters, we assume it will also be meaningful to evaluate the sensitivity of security attributes to variations in model parameters.

In Section 5 we present a case study with numerical results to show how one can use our quantitative analysis of system security and the influences of changes in various model parameters. Here, we can compute the steady-state probabilities $\{\pi_i, i \in X_s\}$ of the SMP states by using Eqs. 4 and 8. Again, for the sake of brevity, we assume:

$\Phi = (p_i p_a + 1 - p_a + p_a p_r) h_I + p_t h_T + p_t p_a h_A + (p_i p_a + 1 - p_a + p_a p_r - p_i p_t p_a) h_R.$
The solutions are presented as

$$\pi_I = \frac{p_i p_a + 1 - p_a + p_a p_r}{\Phi} h_I \tag{9}$$

$$\pi_T = \frac{p_t}{\Phi} h_T \tag{10}$$

$$\pi_A = \frac{p_t p_a}{\Phi} h_A \tag{11}$$

$$\pi_R = \frac{h_R}{h_I} \pi_I - \frac{p_i p_t p_a}{\Phi} h_R \tag{12}$$

Given the SMP model steady-state probabilities, various measures can be computed via Eqs. 1 to 3.

3.3 Sensitivity Analysis

The main aim of parametric sensitivity analysis is to predict the effect of variations in inputs and parameters on outputs (measures), hoping to find performance or reliability bottlenecks, and guiding an optimisation process [20]. It is a useful procedure for offloading system optimisation in the early design phase. Since some model parameters are difficult to ascertain in the design phase, sensitivity analysis can predict the influence on the quantitative analysis results from changes in different parameters. Matos et al. [21] developed a hierarchical analytical model of mobile cloud availability and presented sensitivity analysis based on distinct techniques to assess the impact of each input parameter to identify the bottlenecks for system improvement. Effects on the measures on SMP from changes in different input parameters are discussed.

$Measure \in \{Cost, Confid, Trade\}$ is a measure. $x \in \{p_i, p_t, p_a, p_r, h_I, h_T, h_A, h_R\}$ is a variable in our model. The sensitivity analysis is conducted by calculating the derivative of the measure with respect to a certain input parameter.

$$\frac{d(Measure)}{dx} \tag{13}$$

Eqs. 13 is the sensitivity formula for measure prediction in the SMP model. The numerical results in the format of graphs will be shown in the next section, from which we can see intuitively the impact of parameter changes on different measures.

4 Numerical Study

In this section we give numerical results as examples to show how one can evaluate security attributes of the SMP model defined in the previous sections using different measures.

First, we assume that the probability of a timing attack coming to the offloading system is equal to the one that the system will trigger its rekeying process, i.e., $p_t = 0.5$. The mean time the system spends before an attacker conducts a timing attack or it rekeys is $h_I = 10$ time units. Further, the probability that the attacker successfully cracks the system secret using a timing attack is $p_a = 0.6$ and the probability of an unsuccessful attack $1 - p_a = 0.4$. The time taken by a successful timing attack is assumed to be $h_T = 5$ time unites. Besides, suppose that the probability that the system return initial state by manual intervention is $p_i = 0.2$ and probability of the attack is terminated due to rekeying operation is $p_r = 0.5$. Hence, the probability that the current attack stops and another timing attack affects the system is $1 - p_i - p_r = 0.3$. We also assume the duration for a specific attack is supposed to be $h_A = 3$ time unites and rekeying process time is $h_R = 1$ time unite respectively.

Using the values given above as the model input parameters and Eqs. 9 - 12, we obtain the steady-state probabilities of the Semi-Markov process as:

$\pi_I = 0.6634$, $\pi_T = 0.2023$, $\pi_A = 0.0728$, $\pi_R = 0.0615$.

The steady-state probabilities π_i may be interpreted as the proportion of time that the SMP spends in the state i. For the assumed values of the input parameters, the proportion of time that the offloading system spends in the initial state I is approximately 66% of the whole system life time.

Fig. 2 shows the measure of system cost, confidentiality and trade-off metric changing with different weighting parameters c. To better scale for the figure, the trade-off metric $Trade$ is divided by 15 and the $Cost$ metric is multiplied by 10. From Fig. 2, it can be seen that the system cost increase with rekey probability p_r when the weighting parameter is small, as we put more weight on the rekeying cost. At high values, the information loss cost component becomes the decisive factor. We see the decrease in system cost as p_r increase. While one can see that the cost increases with rekey probability p_r, the confidentiality metric stays content with changing weighting parameters c as it is independent of the weighting. The combined trade-off measure increases as the rekey probability p_r increases when the parameter c is small. All three metrics achieve their minimum value when $c = 1$, where we only consider the costs of the compromised systems. From Fig. 3, we can observe that the system cost metric is very large when h_I is small with all possible weighting parameter values, since the system triggers rekey process not so frequently that the system sojourns in the initial state I for a short time. Again, for all values of the weighting parameter c from 0 to 1, the confidentiality metric stays the same and it increases when h_I becomes larger.

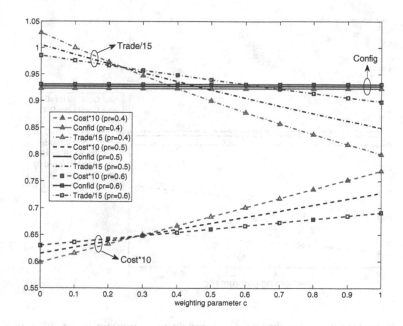

Fig. 2. System measures changing with weighting parameter under different p_r

Intuitively, one can see that the larger the mean sojourn time in state I, the better is the trade-off measure. However, it decreases with increasing weighting parameter c as we put less weight to the system rekeying cost. In the flowing analysis, we choose to put more weight to the cost with system information lost in state A than the cost causes by the rekeying process, as $c = 0.7$.

Fig. 4 shows how different measures behave with changes in rekey probability parameter p_r and the time in the initial state h_I. It can be seen that the larger p_r and h_I are, the better the offloading system performs.

As discussed in Section 2.2, p_r is the probability that the attack is terminated due to rekeying operation, so it depends on the system configuration. And rekeying process can bring the system back to initial state before an attack, which will affect h_I. Since we can tune the rekeying process as system administrators, we conduct sensitivity analysis of system behaviour on the effect from changes in the rekey probability p_r and the mean sojourn time in initial state h_I . Fig. 5(a) shows the system cost measure as a function of h_I and p_r. Interestingly, when the mean time in initial state h_I is short, the system cost increase as the rekey probability p_r increase. However, we see a decrease in system cost as p_r increase, when h_I is very long. Also we can see, the system cost is more sensitive to h_I than to p_r. In Fig. 5(b), we conduct sensitivity analysis to the system confidentiality measure $Confid$. It increase dramatically with the sojourn time h_I in state I when the system is rekeying more frequently. However, it does not interact that strikingly with model input parameter p_r.

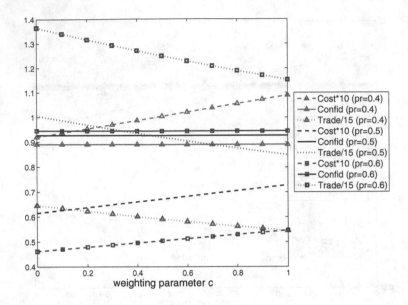

Fig. 3. System measures changing with weighting parameter under different h_I

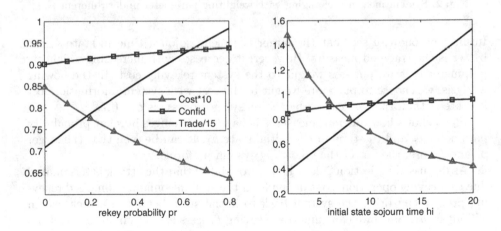

Fig. 4. System measure comparison under p_r and h_I

The trade-off metric as a function of h_I and p_r is depicted in Fig. 6. As expected, the trade-off metric monotonically increase as p_r and h_I increase. That is because the system more often rekeys and it spends more time in good state. When the time in initial state is short, the trade-off measure hardly changes with the parameter p_r. While the rekey probability has a significant effect on the system as h_I is very large. Therefore, a good system management will be able to enlarge the mean sojourn time in initial state.

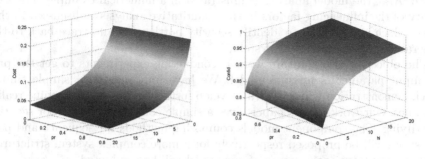

(a) $Cost$ as a function of h_I and p_r (b) $Confid$ as a function of h_I and p_r

Fig. 5. Sensitivity analysis to $Cost$ and $Config$

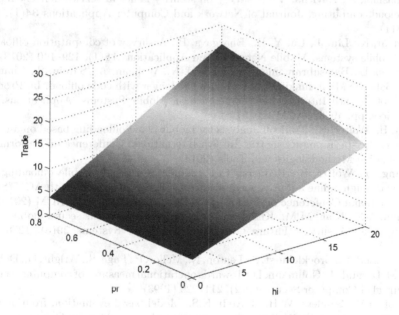

Fig. 6. Trade-off metric as a function of h_I and p_r

5   Conclusion and Future Work

In this paper, we have presented an approach for quantitative assessment of security attributes for an offloading system under the specific threat of timing attacks. A state transition model that describes the dynamic behavior of this system is used as the basics for developing a stochastic model. We have solved for steady-state probabilities of the Semi-Markov Process model as the foundation of security attributes analysis. These include system cost and a trade-off metric. Also, the model analysis is illustrated in a numerical example. The sensitivity of the influencing factors in the quantitative analysis is discussed in this paper, which can be used to identify security bottlenecks and trace back to the vulnerability of the offloading system.

The objective of our future work is conducting experiments to get the precise input parameters for our model. We have implemented a simple timing attack demonstration which we will develop further and use to determine realistic values. Our work in this paper gives a simple component model for security quantifying. The component models composing methods for sequence and parallel styles will be proposed respectively form more complex system structures. Furthermore, transient measure of our model will be conducted.

References

1. Subashini, S., Kavitha, V.: A survey on security issues in service delivery models of cloud computing. Journal of Network and Computer Applications **34**(1), 1–11 (2011)
2. Kumar, K., Liu, J., Lu, Y.-H., Bhargava, B.: A survey of computation offloading for mobile systems. Mobile Networks and Applications **18**(1), 129–140 (2013)
3. Cuervo, E., Balasubramanian, A., Cho, D.-K., Wolman, A., Saroiu, S., Chandra, R., Bahl, P.: Maui: making smartphones last longer with code offload. In: Proceedings of the 8th International Conference on Mobile Systems, Applications, and Services, pp. 49–62. ACM (2010)
4. Wu, H., Wolter, K.: Tradeoff analysis for mobile cloud offloading based on an additive energy-performance metric. In: 8th International Conference on Performance Evaluation Methodologies and Tools (2014)
5. Wang, Q., Wolter, K.: Reducing task completion time in mobile offloading systems through online adaptive local restart. In: Proceedings of the 6th ACM/SPEC International Conference on Performance Engineering, pp. 3–13. ACM (2015)
6. Khan, A.N., Kiah, M.M., Khan, S.U., Madani, S.A.: Towards secure mobile cloud computing: A survey. Future Generation Computer Systems **29**(5), 1278–1299 (2013)
7. Littlewood, B., Brocklehurst, S., Fenton, N., Mellor, P., Page, S., Wright, D., Dobson, J., McDermid, J., Gollmann, D.: Towards operational measures of computer security. Journal of Computer Security **2**(2), 211–229 (1993)
8. Nicol, D.M., Sanders, W.H., Trivedi, K.S.: Model-based evaluation: from dependability to security. IEEE Transactions on Dependable and Secure Computing **1**(1), 48–65 (2004)
9. Zhang, J.-F., Liu, F., Zheng, L.-M., Jia, Y., Zou, P.: Using network security index system to evaluate network security. In: Qi, E., Shen, J., Dou, R. (eds.) The 19th International Conference on Industrial Engineering and Engineering Management, pp. 989–1000. Springer, Heidelberg (2013)
10. Lenkala, S.R., Shetty, S., Xiong, K.: Security risk assessment of cloud carrier. In: 2013 13th IEEE/ACM International Symposium on Cluster, Cloud and Grid Computing (CCGrid), pp. 442–449. IEEE (2013)
11. Rebeiro, C., Mukhopadhyay, D., Bhattacharya, S.: An introduction to timing attacks. In: Timing Channels in Cryptography, pp. 1–11. Springer (2015)
12. Limnios, N., Oprisan, G.: Semi-Markov processes and reliability. Springer Science & Business Media (2001)

13. Köpf, B., Basin, D.: Automatically deriving information-theoretic bounds for adaptive side-channel attacks. Journal of Computer Security **19**(1), 1–31 (2011)
14. Brumley, B.B., Tuveri, N.: Remote timing attacks are still practical. In: Atluri, V., Diaz, C. (eds.) ESORICS 2011. LNCS, vol. 6879, pp. 355–371. Springer, Heidelberg (2011)
15. Brumley, D., Boneh, D.: Remote timing attacks are practical. Computer Networks **48**(5), 701–716 (2005)
16. Weiß, Michael, Heinz, Benedikt, Stumpf, Frederic: A cache timing attack on aes in virtualization environments. In: Keromytis, Angelos D. (ed.) FC 2012. LNCS, vol. 7397, pp. 314–328. Springer, Heidelberg (2012)
17. Palanisamy, B., Liu, L.: Mobimix: protecting location privacy with mix-zones over road networks. In: 2011 IEEE 27th International Conference on Data Engineering (ICDE), pp. 494–505. IEEE (2011)
18. Avizienis, A., Laprie, J.-C., Randell, B., Landwehr, C.: Basic concepts and taxonomy of dependable and secure computing. IEEE Transactions on Dependable and Secure Computing **1**(1), 11–33 (2004)
19. Trivedi, K.S.: Probability & statistics with reliability, queuing and computer science applications. John Wiley & Sons (2008)
20. Frank, P.M.: Introduction to system sensitivity theory, vol. 11. Academic press, New York (1978)
21. Matos, R., Araujo, J., Oliveira, D., Maciel, P., Trivedi, K.: Sensitivity analysis of a hierarchical model of mobile cloud computing. Simulation Modelling Practice and Theory 50, 151–164 (2015). Special Issue on Resource Management in Mobile Clouds

Single-Server Systems with Power-Saving Modes

Tuan Phung-Duc[✉]

Department of Mathematical and Computing Sciences,
Tokyo Institute of Technology, Ookayama, Meguro-ku, Tokyo, Japan
`tuan@is.titech.ac.jp`

Abstract. Vacations queues are motivated from the need of utilizing the server when it is idle. Most of papers in the literature assume that consecutive vacations follow the same distribution. Recently, Ibe et al. [5] consider a model where the lengths of consecutive vacations follow different distributions and obtain the steady state solution by a direct method. In this paper, we first consider the same model and obtain exact results as well as the decomposition for the queue length and the sojourn time via a generating function approach. We then demonstrate that our method can analyze more complex models with working vacations or with abandonment. Numerical results show insights into the performance of single server systems with power-saving modes.

Keywords: Setup time · Working vacation · Abandonment · Power-saving

1 Introduction

Vacation queue is characterized by the feature that the server may be unavailable for service for a random period of time when it is idle. The time that the server is away from service is called vacation. Vacation is resulted from many factors. In some cases, vacation is resulted from post service processing, server breakdowns etc. Some other case, vacation corresponds to a power-saving mode where the server is turned off in order to save energy in communication and computer systems. This is because in the current technology, an idle server still consumes about 60% of its peak consumption [3]. This paper pays attention to power-consumption and thus we refer the vacation to as the power-saving mode.

Two simple vacation policies are single vacation and multiple vacation. In the former, the server takes one vacation and returns to normal mode even if there is no customer in the system. In the latter case, upon the completion of a vacation, if the system is empty, the server takes another vacation otherwise it starts serving waiting customers. This paper focuses on the latter case, i.e., the server takes vacations until it finds a waiting customer upon completion of a vacation. In the literature, most of papers deal with the case where vacations are homogeneous, i.e. consecutive vacations follow the same distribution. Recently, Ibe et al. [5] analyzes a model in which the distribution of the duration of the first vacation is different from that of other vacations. The authors obtain

© Springer International Publishing Switzerland 2015
M. Gribaudo et al. (Eds.): ASMTA 2015, LNCS 9081, pp. 158–172, 2015.
DOI: 10.1007/978-3-319-18579-8_12

the steady state result using a simple method based on difference equations. A related work is due to Ke [7] where two type of vacations based on thresholds are presented. The notation of differentiated vacations are also introduced in [1,14] in the framework of gated vacation while the multiple adaptive vacations are presented in [13] in a discrete time context.

In this paper, using a generating function approach, we give a simpler solution for the model by Ibe et al. [5] and some extended models. In particular, we first analyze the model by Ibe et al. [5] using generating function approach which yields simple expressions. We obtain the generating function for the number of customers in the system based on which the sojourn time distribution is easily obtained via the distributional Little's law [8]. We then analyze a more general model with differentiated working vacations. This model generalizes those proposed by Servi and Finn [12] and Ibe et al. [5]. Finally, we analyze an extension of the model of [5] where customers may abandon the service when the server is on vacation. In addition, deep insights such as decomposition properties are easily obtained with the generating function methodology. It should be noted that in single server context, the model with multiple vacation is identical to that with setup time where the first customer of each busy period should wait for a setup time. Queues with setup time are extensively studied recently because they have application in power-saving data centers [2,4,6,9–11].

The rest of this paper is presented as follows. Section 2 analyzes the model presented by Ibe et al. [5] using a generating function approach. Section 3 deals with an extended model with working vacation. Section 4 presents the analysis of the model with abandonment. Some insights into the performance of these system are presented in Section 5 via numerical experiments. Finally, concluding remarks are presented in 6.

2   Model with Differentiated Vacations

In this section, we describe the M/M/1 model with differentiated vacations [5] and present a new analysis via generating functions.

2.1   Model

We consider an M/M/1 queueing system with vacation. Customers arrive at the system according a Poisson process with rate λ and request for an exponentially distributed service with mean $1/\mu$. The server starts a type I vacation when it becomes idle. The length of this vacation follows an exponential distribution with mean $1/\gamma_1$. On returning from this vacation, if the server is still idle, it takes a type II vacation whose duration follows an exponential distribution with mean $1/\gamma_2$. Type II vacations are repeated as long as the system is empty upon the completion of a vacation. On returning from either a type I or type II vacation, if there are some customers in the system the server immediately starts servicing customers until the system is empty again.

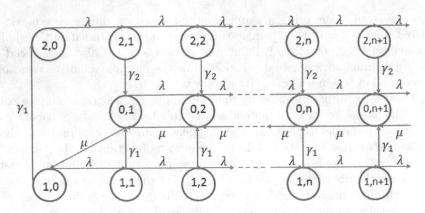

Fig. 1. Model with differentiated vacations

2.2   Analysis

Let $S(t)$ denote the state of the server,

$$S(t) = \begin{cases} 0, & \text{serving,} \\ 1, & \text{vacation of type I,} \\ 2, & \text{vacation of type II.} \end{cases}$$

Let $N(t)$ denote the number of customers in the system. It is easy to see that $(S(t), N(t))$ forms a Markov chain on state space

$$\mathcal{S} = \{(0,j); j \in \mathbb{N}\} \cup \{(1,j); j \in \mathbb{Z}_+\} \cup \{(2,j); j \in \mathbb{Z}_+\},$$

where $\mathbb{N} = \{1, 2, \dots\}$ and $\mathbb{Z}_+ = 0 \cup \mathbb{N}$. See Figure 1 for the transitions among states.

Let $\pi_{i,j}$ denote the steady state probability that the system is in state (i, j). Balance equations for state $(1, j), j \in \mathbb{Z}_+$ are given by

$$(\lambda + \gamma_1)\pi_{1,j} = \lambda \pi_{1,j-1}, \qquad j \geq 1. \tag{1}$$

Let $\Pi_1(z) = \sum_{j=0}^{\infty} \pi_{1,j} z^j$ denote the generating function of $\pi_{1,j}, j \in \mathbb{Z}_+$. Multiplying (1) by z^j ($j \in \mathbb{N}$), summing up over $j \in \mathbb{N}$ and arranging the result yields

$$(\lambda + \gamma_1)(\Pi_1(z) - \pi_{1,0}) = \lambda z \Pi_1(z),$$

leading to

$$\Pi_1(z) = \frac{(\lambda + \gamma_1)\pi_{1,0}}{\lambda + \gamma_1 - \lambda z}.$$

Similarly, we also have

$$\Pi_2(z) = \frac{(\lambda + \gamma_2)\pi_{2,0}}{\lambda + \gamma_2 - \lambda z},$$

where $\Pi_2(z) = \sum_{j=0}^{\infty} \pi_{2,j} z^j$ denotes the generating function of $\pi_{2,j}$ ($j \in \mathbb{Z}_+$).

Balance equations for state $(0, j), j \in \mathbb{N}$ are given by.

$$(\lambda + \mu)\pi_{0,1} = \mu\pi_{0,2} + \gamma_1\pi_{1,1} + \gamma_2\pi_{2,1}, \tag{2}$$
$$(\lambda + \mu)\pi_{0,j} = \lambda\pi_{0,j-1} + \mu\pi_{0,j+1} + \gamma_1\pi_{1,j} + \gamma_2\pi_{2,j}, \qquad j \geq 2. \tag{3}$$

Let $\Pi_0(z) = \sum_{j=1}^{\infty} \pi_{0,j} z^j$ denote the generating function of $\pi_{0,j}$ ($j \in \mathbb{N}$). Multiplying (2) by z and (3) by z^j and summing up the results over $j \in \mathbb{N}$, we obtain

$$(\lambda+\mu)\Pi_0(z) = \lambda z \Pi_0(z) + \frac{\mu}{z}(\Pi_0(z) - \pi_{0,1}z) + \gamma_1(\Pi_1(z) - \pi_{1,0}) + \gamma_2(\Pi_2(z) - \pi_{2,0}).$$

leading to

$$[(\lambda + \mu)z - \lambda z^2 - \mu]\Pi_0(z) = -\mu\pi_{0,1}z + \gamma_1 z(\Pi_1(z) - \pi_{1,0}) + \gamma_2 z(\Pi_2(z) - \pi_{2,0}).$$

Substituting the generating functions $\Pi_1(z)$ and $\Pi_2(z)$ into the above equation and arranging the result yields,

$$\Pi_0(z) = \frac{\lambda z}{\mu - \lambda z}(\Pi_1(z) + \Pi_2(z)).$$

Let $\Pi(z)$ denote the generating function of the number of customers in the system. We have

$$\Pi(z) - \Pi_0(z) + \Pi_1(z) + \Pi_2(z) = \frac{\mu}{\mu - \lambda z}(\Pi_1(z) + \Pi_2(z))$$
$$= \frac{1 - \rho}{1 - \rho z} \cdot \frac{\Pi_1(z) + \Pi_2(z)}{1 - \rho},$$

where $\rho = \lambda/\mu$. Thus, from the normalization condition, we obtain $\Pi_1(1) + \Pi_2(1) = 1 - \rho$. Balance equation for state $(2, 0)$ yields, $\lambda\pi_{2,0} = \gamma_1\pi_{1,0}$. From these two equations, we obtain

$$\pi_{1,0} = \frac{1 - \rho}{\frac{\lambda+\gamma_1}{\gamma_1} + \frac{\gamma_1}{\lambda}\frac{\lambda+\gamma_2}{\gamma_2}}, \qquad \pi_{2,0} = \frac{(1 - \rho)\frac{\gamma_1}{\lambda}}{\frac{\lambda+\gamma_1}{\gamma_1} + \frac{\gamma_1}{\lambda}\frac{\lambda+\gamma_2}{\gamma_2}}.$$

Furthermore, we have

$$\pi_{0,1} = \frac{1 - \rho}{\frac{\mu}{\gamma_1} + \frac{\mu\gamma_1(\lambda+\gamma_2)}{\lambda\gamma_2(\lambda+\gamma_1)}}.$$

Remark 1. It should be noted that if $\gamma_2 \to \infty$, the system converges to the M/M/1 queue with single vacation. In fact, we have

$$\lim_{\gamma_2 \to \infty} \pi_{1,0} = \frac{1 - \rho}{\frac{\lambda+\gamma_1}{\gamma_1} + \frac{\gamma_1}{\lambda}}, \qquad \lim_{\gamma_2 \to \infty} \pi_{2,0} = \frac{(1 - \rho)\frac{\gamma_1}{\lambda}}{\frac{\lambda+\gamma_1}{\gamma_1} + \frac{\gamma_1}{\lambda}}, \qquad \lim_{\gamma_2 \to \infty} \Pi_2(z) = \pi_{2,0}.$$

State $(2, 0)$ corresponds to the idle state in the corresponding M/M/1 queue with single vacation.

Remark 2. Since

$$\lim_{\gamma_1 \to \infty} \pi_{1,0} = 0, \qquad \lim_{\gamma_1 \to \infty} \pi_{2,0} = \frac{(1-\rho)\gamma_2}{\lambda + \gamma_2},$$

the current model tends to the M/M/1 with multiple vacations as $\gamma_1 \to \infty$.

We rewrite the generating function of the number of customers in the system as follows.

$$\begin{aligned}
\Pi(z) &= \frac{1-\rho}{1-\rho z} \frac{\Pi_1(z) + \Pi_2(z)}{1-\rho} \\
&= \frac{1-\rho}{1-\rho z} \left(\frac{\Pi_1(1)}{1-\rho} \frac{\gamma_1}{\lambda + \gamma_1 - \lambda z} + \frac{\Pi_2(1)}{1-\rho} \frac{\gamma_2}{\lambda + \gamma_2 - \lambda z} \right).
\end{aligned}$$

The first term in the right hand side is the generating function of the number of customers in an M/M/1 queue while the second term corresponds to the number of customers that arrive during the remaining time of vacations. More precisely, $\frac{\Pi_1(1)}{1-\rho}$ and $\frac{\Pi_2(1)}{1-\rho}$ are the probabilities that the server is in vacation of type I and II respectively under the condition that the server is not working.

Furthermore, it should be noted that the distributional Little's law [8] establishes in our model. Therefore, the LST of the sojourn time distribution is given by

$$W^*(s) = \Pi(1 - \frac{s}{\lambda}),$$

which is also decomposed into two parts. The first one corresponds to the sojourn time in the conventional M/M/1 queue without vacation while the second one corresponds to the extra sojourn time due to vacations. More precisely, the LST of the sojourn time is given by

$$W^*(s) = \frac{\mu - \lambda}{\mu - \lambda + s} \left(\frac{\Pi_1(1)}{1-\rho} \frac{\gamma_1}{\gamma_1 + s} + \frac{\Pi_2(1)}{1-\rho} \frac{\gamma_2}{\gamma_2 + s} \right).$$

The mean number of customers in the system $\mathrm{E}[L]$ is given by $\Pi'(1)$, i.e.,

$$\mathrm{E}[L] = \Pi'(1) = \frac{\rho}{1-\rho} + \frac{\Pi_1(1)}{1-\rho} \frac{\lambda}{\gamma_1} + \frac{\Pi_2(1)}{1-\rho} \frac{\lambda}{\gamma_2}.$$

The mean sojourn time $\mathrm{E}[W]$ is obtained using the Little's law

$$\mathrm{E}[W] = \frac{\Pi'(1)}{\lambda}.$$

3   Model with Differentiated Working Vacations

In this section, we analyze a model with differentiated working vacations. In this model the server still processes job but at different rates in vacations.

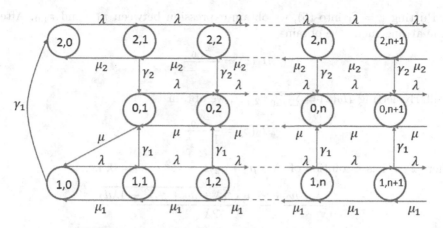

Fig. 2. Model with differentiated working vacations

3.1 Model

We further consider the case where in vacation state the server can still serves customers but with different rate. In particular, we assume that the service rates at type I and type II vacations are μ_1 and μ_2, respectively. Since the state space is the same as that of the model in Section 2, we use the same notation for the stationary distribution. See Figure 2 for the transition among states.

3.2 Analysis

Balance equation is now more complex. Balance equations for states $(1, j), j \in \mathbb{N}$ are given by

$$(\lambda + \mu_1 + \gamma_1)\pi_{1,1} = \lambda\pi_{1,0} + \mu_1\pi_{1,2}, \qquad j = 1, \tag{4}$$

$$(\lambda + \mu_1 + \gamma_1)\pi_{1,j} = \lambda\pi_{1,j-1} + \mu_1\pi_{1,j+1}, \qquad j \geq 2. \tag{5}$$

Let $\Pi_1(z) = \sum_{j=0}^{\infty} \pi_{1,j}z^j$ denote the generating function of $\pi_{1,j}$ $(j \in \mathbb{Z}_+)$. Multiplying (4) by z and (5) by z^j, summing up over $j \in \mathbb{N}$ and arranging the result, we obtain

$$(\lambda + \mu_1 + \gamma_1)(\Pi_1(z) - \pi_{1,0}) = \lambda z \Pi_1(z) + \frac{\mu_1}{z}(\Pi_1(z) - \pi_{1,0} - \pi_{1,1}z),$$

leading to

$$\Pi_1(z) = \frac{[(\lambda + \mu_1 + \gamma_1)\pi_{1,0} - \mu_1\pi_{1,1}]z - \mu_1\pi_{1,0}}{(\lambda + \mu_1 + \gamma_1)z - \lambda z^2 - \mu_1}. \tag{6}$$

Let $0 < \widehat{z}_1 < 1$ and $z_1 > 1$ denote two distinct roots of the denominator of the above formula. We have

$$\widehat{z}_1 = \frac{\lambda + \mu_1 + \gamma_1 - \sqrt{(\lambda + \mu_1 + \gamma_1)^2 - 4\lambda\mu_1}}{2\lambda},$$

$$z_1 = \frac{\lambda + \mu_1 + \gamma_1 + \sqrt{(\lambda + \mu_1 + \gamma_1)^2 - 4\lambda\mu_1}}{2\lambda}.$$

Putting $z = \hat{z}_1$ into (6), we obtain expression between $\pi_{1,1}$ and $\pi_{1,0}$. After some arrangement, we obtain

$$\Pi_1(z) = \frac{\pi_{1,0}}{1 - \frac{z}{z_1}}.$$

Similarly, letting $\Pi_2(z) = \sum_{j=0}^{\infty} \pi_{2,j} z^j$, we obtain

$$\Pi_2(z) = \frac{\pi_{2,0}}{1 - \frac{z}{z_2}},$$

where $z_2 > 1$ is a solution of $(\lambda + \mu_2 + \gamma_2)z - \lambda z^2 - \mu_2 = 0$, i.e.,

$$z_2 = \frac{\lambda + \mu_2 + \gamma_2 + \sqrt{(\lambda + \mu_2 + \gamma_2)^2 - 4\lambda\mu_2}}{2\lambda}.$$

We write down the balance equations for states $(0,j)$ $(j \in \mathbb{N})$.

$$(\lambda + \mu)\pi_{0,1} = \mu\pi_{0,2} + \gamma_1\pi_{1,1} + \gamma_2\pi_{2,1}, \qquad j = 1, \tag{7}$$
$$(\lambda + \mu)\pi_{0,j} = \lambda\pi_{0,j-1} + \mu\pi_{0,j+1} + \gamma_1\pi_{1,j} + \gamma_2\pi_{2,j}, \qquad j \geq 2. \tag{8}$$

Let $\Pi_0(z) = \sum_{j=1}^{\infty} \pi_{0,j} z^j$ denote the generating function of $\pi_{0,j}$ $(j \in \mathbb{N})$. Multiplying (7) by z and (8) by z^j and summing up the result over $j \in \mathbb{N}$ yields

$$(\lambda + \mu)\Pi_0(z) = \lambda z \Pi_0(z) + \frac{\mu}{z}(\Pi_0(z) - \pi_{0,1}z) + \gamma_1\Pi_1(z) + \gamma_2\Pi_2(z).$$

Arranging this equation, we obtain

$$(z-1)(\mu - \lambda z)\Pi_0(z) = -\mu\pi_{0,1}z + \gamma_1 z(\Pi_1(z) - \pi_{1,0}) + \gamma_2 z(\Pi_2(z) - \pi_{2,0})$$
$$= \gamma_1 z(\Pi_1(z) - \Pi_1(1)) + \gamma_2 z(\Pi_2(z) - \Pi_2(1)).$$

Simplifying this expression we obtain

$$\Pi_0(z) = \frac{z}{\mu - \lambda z}\left(\frac{\gamma_1\Pi_1(z)}{z_1 - 1} + \frac{\gamma_2\Pi_2(z)}{z_2 - 1}\right).$$

Thus we have

$$\Pi(z) = \sum_{i=0}^{2} \Pi_i(z)$$

$$= \left(\frac{\gamma_1 z}{(\mu - \lambda z)(z_1 - 1)} + 1\right)\Pi_1(z) + \left(\frac{\gamma_2 z}{(\mu - \lambda z)(z_2 - 1)} + 1\right)\Pi_2(z).$$

Furthermore, using a cut between $\{\pi_{1,j}; j \in \mathbb{Z}_+\}$ and $\{\pi_{2,j}; j \in \mathbb{Z}_+\}$ yields

$$\gamma_1\pi_{1,0} = \gamma_2(\Pi_2(1) - \pi_{2,0}),$$

or equivalently

$$\pi_{2,0} = \frac{(z_2 - 1)\gamma_1}{\gamma_2}\pi_{1,0}.$$

Using the normalization condition $\Pi(1) = 1$, we obtain $\pi_{1,0}$.

Remark 3. We observe the following limiting results.

$$\lim_{\gamma_2 \to \infty} z_2 = \infty, \qquad \lim_{\gamma_2 \to \infty} \Pi_2(z) = \pi_{2,0}, \qquad \lim_{\gamma_2 \to \infty} \pi_{2,0} = \lim_{\gamma_2 \to \infty} \frac{\gamma_1}{\lambda} \pi_{1,0}.$$

Thus when $\gamma_2 \to \infty$, our model reduces to the M/M/1 model with single working vacation. In this limiting regime, state $(2,0)$ is the idle state in the corresponding M/M/1 model with single working vacation.

Remark 4. We have

$$\lim_{\gamma_1 \to \infty} z_1 = \infty, \qquad \lim_{\gamma_1 \to \infty} \pi_{1,0} = 0, \qquad \lim_{\gamma_1 \to \infty} \Pi_1(z) = 0.$$

Thus, when $\gamma_1 \to \infty$, our model reduces to the M/M/1 queue with working vacation (M/M/1/WV) [12].

For this model, we could also write the generating function $\Pi(z)$ in the following form.

$$\Pi(z) = \frac{1 - \rho}{1 - \rho z} \Pi_v(z),$$

where $\Pi_v(z)$ is the generating function of the extra queue length inducing by working vacations. However, the form of $\Pi_v(z)$ is more involved and does not have a clear physical interpretation as in the model presented in Section 2.

Furthermore, it should be noted that the distributional Little's law [8] establishes for our model. Thus, the Laplace-Stieltjes transform (LST) of the sojourn time distribution is given by

$$W^*(s) = \Pi(1 - \frac{s}{\lambda}),$$

which is also decomposed into two parts. The first one corresponds to the sojourn time in the M/M/1 queue without vacation while the second one corresponds to the extra sojourn time due to working vacations.

Using the LST of the sojourn time distribution, we derive the mean and the variance of the sojourn time in the system. We have

$$\Pi_1'(1) = \frac{\pi_{1,0} z_1}{(z_1 - 1)^2}, \qquad \Pi_2'(1) = \frac{\pi_{2,0} z_2}{(z_2 - 1)^2}$$

$$\Pi_0'(1) = \frac{\mu}{(\mu - \lambda)^2} \left(\frac{\gamma_1 \Pi_1(1)}{z_1 - 1} + \frac{\gamma_2 \Pi_2(1)}{z_2 - 1} \right) + \frac{1}{\mu - \lambda} \left(\frac{\gamma_1 \Pi_1'(1)}{z_1 - 1} + \frac{\gamma_2 \Pi_2'(1)}{z_2 - 1} \right).$$

The mean number of customers in the system $E[L]$ is given by $\Pi'(1)$.

$$E[L] = \Pi'(1) = \Pi_0'(1) + \Pi_1'(1) + \Pi_2'(1).$$

Let W denote the the sojourn time of a customer, we have

$$E[W] = -\left. \frac{dW^*(s)}{ds} \right|_{s=0} = \frac{\Pi'(1)}{\lambda} = \frac{\Pi_0'(1) + \Pi_1'(1) + \Pi_2'(1)}{\lambda},$$

$$E[W^2] = \left. \frac{d^2 W^*(s)}{ds^2} \right|_{s=0} = \frac{\Pi''(1)}{\lambda^2} = \frac{\Pi_0''(1) + \Pi_1''(1) + \Pi_2''(1)}{\lambda^2},$$

where

$$\Pi_1''(1) = \frac{2\pi_{1,0}z_1}{(z_1 - 1)^3}, \qquad \Pi_2''(1) = \frac{2\pi_{2,0}z_2}{(z_2 - 1)^3},$$

$$\Pi_0''(1) = \frac{2\lambda\mu}{(\mu - \lambda)^3}\left(\frac{\gamma_1\Pi_1(1)}{z_1 - 1} + \frac{\gamma_2\Pi_2(1)}{z_2 - 1}\right) + \frac{2\mu}{(\mu - \lambda)^2}\left(\frac{\gamma_1\Pi_1'(1)}{z_1 - 1} + \frac{\gamma_2\Pi_2'(1)}{z_2 - 1}\right)$$
$$+ \frac{1}{\mu - \lambda}\left(\frac{\gamma_1\Pi_1''(1)}{z_1 - 1} + \frac{\gamma_2\Pi_2''(1)}{z_2 - 1}\right).$$

Furthermore, the variance of the sojourn time is given by

$$\mathrm{Var}[W] = \mathrm{E}[W^2] - \mathrm{E}[W]^2.$$

Thus, we have explicit expressions for the mean, the variance as well as the LST of the sojourn time distribution.

4   Model with Abandonment

In this section, we analyze a variant of the model in Section 2, where customers may abandon when the server is on vacations.

4.1   Model

In this section, we investigate the model in Section 2 adding the feature that customers may abandon when the server is on vacations. In this model, customers may abandon after some exponentially distributed waiting time with mean $1/\theta$ in type I vacation. If the server is in type II vacation, customers abandon after an exponentially distributed waiting time with mean $1/\varphi$. Since the state space is the same as before, we use the same notations for the stationary distribution.

4.2   Analysis

Balance equations for states $\{(1, j); j \in \mathbb{Z}_+\}$ are given by

$$(\lambda + \gamma_1)\pi_{0,1} = \mu\pi_{0,1} + \theta\pi_{1,1},$$
$$(\lambda + \gamma_1 + j\theta)\pi_{1,j} = \lambda\pi_{1,j-1} + (j+1)\theta\pi_{1,j+1}.$$

Defining generating function by $\Pi_1(z) = \sum_{j=0}^{\infty}\pi_{0,j}z^j$, we have

$$(\lambda + \gamma_1)\Pi_1(z) + \theta z\Pi_1'(z) = \mu\pi_{0,1} + \lambda z\Pi_1(z) + \theta\Pi_1'(z),$$

which is transformed to

$$\Pi_1'(z) = \frac{\lambda z - \lambda - \gamma_1}{\theta(z - 1)}\Pi_1(z) + \frac{\mu\pi_{0,1}}{\theta(z - 1)}.$$

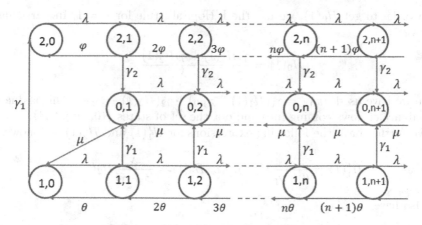

Fig. 3. Model with impatient customer inducing by vacations

After some tedious algebras taking into account the fact that $\Pi_1(1)$ is finite, we obtain

$$\Pi_1(z) = \frac{\mu \pi_{0,1}}{\theta} \exp\left(\frac{\lambda}{\theta}z\right)(1-z)^{-\frac{\gamma_1}{\theta}} \int_z^1 \exp\left(-\frac{\lambda}{\theta}u\right)(1-u)^{\frac{\gamma_1}{\theta}-1}du.$$

Balance equations for states $\{(2,j); j \in \mathbb{Z}_+\}$ are given by

$$\lambda\pi_{2,0} = \gamma_1\pi_{1,0} + \varphi\pi_{2,1},$$
$$(\lambda + \gamma_2 + j\varphi)\pi_{2,j} = \lambda\pi_{2,j-1} + (j+1)\varphi\pi_{2,j+1}.$$

Letting $\Pi_2(z) = \sum_{j=0}^{\infty} \pi_{2,j}z^j$ and transforming these two equations to z-domain, we obtain the following differential equation.

$$\Pi_2'(z) = \left(\frac{\lambda}{\varphi} + \frac{\gamma_2}{\varphi(1-z)}\right)\Pi_2(z) + \frac{\gamma_2\pi_{2,0} + \gamma_1\pi_{1,0}}{\varphi(z-1)}$$

This differential equation is similar to that of $\Pi_1(z)$ and the solution is given by

$$\Pi_2(z) = \frac{\gamma_1\pi_{1,0} + \gamma_2\pi_{2,0}}{\varphi} \exp\left(\frac{\lambda}{\varphi}z\right)(1-z)^{-\frac{\gamma_2}{\varphi}} \int_z^1 \exp\left(-\frac{\lambda}{\varphi}u\right)(1-u)^{\frac{\gamma_2}{\varphi}-1}du.$$

Finally, balance equations for states $\{(0,j); j \in \mathbb{N}\}$ are given as follows.

$$(\lambda + \mu)\pi_{0,1} = \mu\pi_{0,2} + \gamma_1\pi_{1,1} + \gamma_2\pi_{1,2},$$
$$(\lambda + \mu)\pi_{0,j} = \lambda\pi_{0,j-1} + \mu\pi_{0,j+1} + \gamma_1\pi_{1,j} + \gamma_2\pi_{2,j}.$$

Letting $\Pi_0(z) = \sum_{j=1}^{\infty} \pi_{0,j}z^j$, we then obtain

$$\Pi_0(z) = \frac{-\mu\pi_{0,1}z + \gamma_1 z(\Pi_1(z) - \pi_{1,0}) + \gamma_2 z(\Pi_2(z) - \pi_{2,0})}{(\lambda + \mu)z - \lambda z^2 - \mu}.$$

In order to get $\Pi_0(1)$, we use the L'Hopital's rule for $z = 1$. In particular, we have

$$\Pi_0(1) = \frac{\gamma_1 \Pi_1'(1) + \gamma_2 \Pi_2'(1)}{\mu - \lambda},$$

where we have used $\mu\pi_{0,1} = \gamma_1(\Pi_1(1) - \pi_{1,0}) + \gamma_2(\Pi_2(1) - \pi_{2,0})$. This is due to the balance of flows coming into and out the set of states $\{(0,j); j \in \mathbb{N}\}$.

We further have the following expressions for $\Pi_1'(1)$ and $\Pi_2'(1)$ as follows.

$$\Pi_1'(1) = \frac{\lambda}{\theta + \gamma_1}\Pi_1(1), \qquad \Pi_2'(1) = \frac{\lambda}{\varphi + \gamma_2}\Pi_2(1).$$

We also have

$$\Pi_1(1) = \frac{\mu\pi_{0,1}}{\gamma_1}, \qquad \Pi_2(1) = \frac{\gamma_1\pi_{1,0} + \gamma_2\pi_{2,0}}{\gamma_2}.$$

Furthermore, substituting $z = 0$ into the expressions for $\Pi_1(z)$ and $\Pi_2(z)$, we obtain

$$\pi_{1,0} = \frac{\mu\pi_{0,1}}{\theta}\kappa_1, \qquad \pi_{2,0} = \frac{\gamma_1\pi_{1,0} + \gamma_2\pi_{2,0}}{\varphi}\kappa_2,$$

where

$$\kappa_1 = \int_0^1 \exp\left(-\frac{\lambda}{\theta}u\right)(1-u)^{\frac{\gamma_1}{\theta}-1}du, \qquad \kappa_2 = \int_0^1 \exp\left(-\frac{\lambda}{\varphi}u\right)(1-u)^{\frac{\gamma_2}{\varphi}-1}du.$$

Thus, $\Pi_0(1), \Pi_1(1)$ and $\Pi_2(1)$ are expressed in terms of $\pi_{0,1}$ which is uniquely determined using the normalization condition.

$$\Pi_0(1) + \Pi_1(1) + \Pi_2(1) = 1.$$

5   Numerical Examples

In this section we investigate the effect of parameters on performance measures for the model presented in Section 3. We fixed some parameters as follows: $\gamma_2 = 1, \mu = 1, \mu_2 = 0.1$ and $\mu_1 = 0.05$. Figures 4 to 7 show the server state probabilities: $\Pi_0(1)$ (normal mode), $\Pi_1(1)$ (type I vacation) and $\Pi_2(1)$ (type II vacation) against the traffic intensity for the case $\gamma_1 = 1, 0.5, 0.25$ and 0.1, respectively. We observe that $\Pi_2(1)$ decreases with ρ while $\Pi_0(1)$ increases with ρ. Furthermore, we observe that $\Pi_1(1)$ increases and then decreases with $\rho \in (0,1)$. Figures 8 and 9 represent the mean and the variance of the response time for $\gamma_1 = 1, 0.5, 0.25$ and 0.1. We observe that the mean and the variance of the sojourn time decrease with γ_1. This is because a small γ_1 means that the type I vacation period is long leading to a large sojourn time.

We have a close look at the steady probabilities of power-saving states, i.e., $(1,0)$ and $(2,0)$ where the power consumption is low. We fix $\gamma_1 = 1$ and vary $\gamma_2 = 1, 0.5, 0.25, 0.1$ and 0.05. In this scenario, the duration of type I vacation is shorter than that of type II vacation. Other parameters (μ, μ_1, μ_2) are kept the same as before. We observe from Figure 10 that $\pi_{1,0}$ increases with a relatively small ρ and then decreases with a relatively large ρ. On the other hand, we observe from Figure 11 that probability $\pi_{2,0}$ decreases with ρ for all γ_2.

In states $(1,0)$ and $(2,0)$, the server consumes only a small amount of energy. Thus, when the probabilities of these states are large, it is suitable for applying the current model because we can save energy. On the other hand, when the steady state probabilities of states $(1,0)$ and $(2,0)$ are small, there is less chance to save energy. As a result, we may use the ON/IDLE model in this case. We also observe that when the traffic intensity is low, $\pi_{2,0}$ is close to one. This is intuitively true because at low traffic intensity there are no customers in the system in almost all the time and thus the server likely to repeat its multiple power-saving modes, i.e., state $(2,0)$.

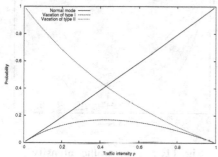

Fig. 4. Probability vs. ρ $(\gamma_1 = 1)$

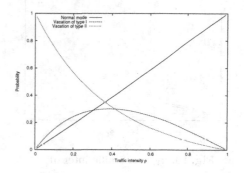

Fig. 5. Probability vs. ρ $(\gamma_1 = 0.5)$

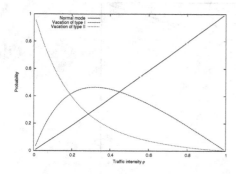

Fig. 6. Probability vs. ρ $(\gamma_1 = 0.25)$

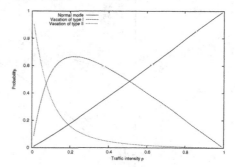

Fig. 7. Probability vs. ρ $(\gamma_1 = 0.1)$

The mean power consumption is given as follows.

$$E[P] = C_0 \Pi_0(1) + C_{10}\pi_{1,0} + C_1(\Pi_1(1) - \pi_{1,0}) + C_2(\Pi_2(1) - \pi_{2,0}) + C_{20}\pi_{2,0}.$$

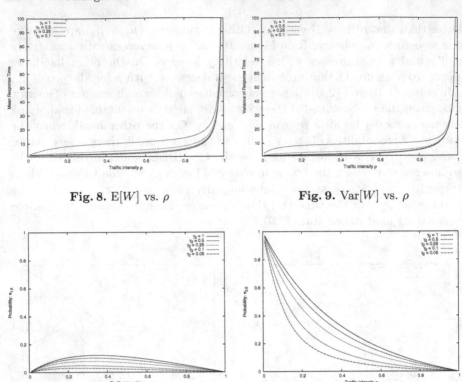

Fig. 8. E[W] vs. ρ

Fig. 9. Var[W] vs. ρ

Fig. 10. $\pi_{1,0}$ vs. traffic intensity

Fig. 11. $\pi_{2,0}$ vs. traffic intensity

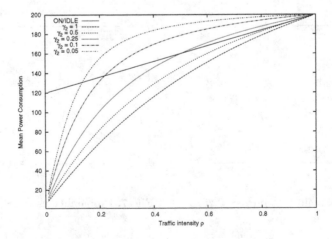

Fig. 12. Power consumption vs. traffic intensity

The coefficients C_0, C_1 and C_2 are the power consumption of the server at normal mode ($\{\pi_{0,j}; j \in \mathbb{N}\}$), type I vacation ($\{\pi_{1,j}; j \in \mathbb{N}\}$) and type II vacation ($\{\pi_{2,j}; j \in \mathbb{N}\}$), respectively. Furthermore, C_{10} and C_{20} are the power consumption of the server at state $(1,0)$ and $(2,0)$ respectively. It should be noted that the server typically consumes only small amount of energy at these states.

Figure 12 shows the power consumption against the traffic intensity. We assume that the server at normal mode consumes $C_0 = 200W$. The power consumption of the server in type I setup is $C_1 = 200W$ while that in type II setup is $C_2 = 210W$. This implies that type II setup is more power-consuming than type I setup. Furthermore, when the server at power saving modes, it consumes only $C_{10} = 15W$ at state $(1,0)$ (suspend state) and $C_{20} = 5W$ at state $(2,0)$ (hibernate), respectively. These data are measured from real experiments [2,6]. Other parameters are fixed as follows: $\gamma_1 = 1, \mu_1 = 0.01, \mu_2 = 0.1, \mu = 1$ and $\gamma_2 = 1, 0.5, 0.25, 0.1, 0.05$. For comparison, we also plot the power consumption per a unit time for the corresponding ON/IDLE model, where we assume that a busy server consumes $200W$ while an idle server still consumes 60% of its peak, i.e, $120W$ [6]. Thus the mean power consumption in the ON/IDLE model is given by $200\rho + 120(1 - \rho)$.

We observe that the curves for $\gamma_2 = 1, 0.5$ are below that for the ON/IDLE model. It means that the current model is always more power-saving than the ON/IDLE model if the setup time of mode II is fast enough. For the cases where $\gamma_2 = 0.25, 0.1, 0.05$, we see that when the traffic intensity is small enough the current model is more power-saving while it is better to adopt ON/IDLE model if the traffic intensity is large enough.

6   Concluding Remark

In this paper, we analyze single queueing systems with differentiated non-working and working vacations. Using the generating function approach, we obtain the generating functions for the queue length distribution which gives exact expression for steady-state probabilities, decomposition property and the LST of the sojourn time distribution. Our models generalize various existing ones in the literature such as models with single or multiple vacations. The methodology in this paper can be extended to the model with more than two working vacation modes.

Acknowledgments. This work was supported in part by JSPS KAKENHI Grant Number 2673001. The author would like to thank the referees for constructive comments which improve the presentation of the paper.

References

1. Fiems, D., Walraevens, J., Bruneel, H.: The discrete-time gated vacation queue revisited. AEU-International Journal of Electronics and Communications **58**(2), 136–141 (2004)

2. Gebrehiwot, M.E., Aalto, S., Lassila, P.: Optimal sleep-state control of energy-aware M/G/1 queues. In: Proc. of 8th International Conference on Performance Evaluation Methodologies and Tools (Valuetools 2014) (2014)
3. Gandhi, A., Harchol-Balter, M., Kozuch, M.A.: Are sleep states effective in data centers? In: Proc. of International Green Computing Conference (IGCC). IEEE (2012)
4. Haverkort, B.R., Postema, B.: Towards simple models for energy-performance trade-offs in data centers. In: Proc. of SOCNET & PGENET 2014 (2014)
5. Ibe, O.C., Isijola, O.A.: M/M/1 multiple vacation queueing systems with differentiated vacations. Modelling and Simulation in Engineering **2014**, Article ID 158247, 6 (2014)
6. Isci, C., McIntosh, S., Kephart, J., Das, R., Hanson, J., Piper, S., Frissora, M.: Agile, efficient virtualization power management with low-latency server power states. In: ACM SIGARCH Computer Architecture News, vol. 41, No. 3, pp. 96–107. ACM (2013)
7. Ke, J.C.: The optimal control of an M/G/1 queueing system with server startup and two vacation types. Applied Mathematical Modelling **27**, 437–450 (2003)
8. Keilson, J., Servi, L.D.: A distributional form of Little's law. Operations Research Letters **7**(5), 223–227 (1988)
9. Phung-Duc, T.: Impatient customers in power-saving data centers. In: Sericola, B., Telek, M., Horváth, G. (eds.) ASMTA 2014. LNCS, vol. 8499, pp. 185–199. Springer, Heidelberg (2014)
10. Phung-Duc, T.: Server farms with batch arrival and staggered setup. In: Proceedings of the Fifth Symposium on Information and Communication Technology, pp. 240–247. ACM (2014)
11. Phung-Duc, T.: Exact solution for M/M/c/Setup queue (2014). http://arxiv.org/abs/1406.3084
12. Servi, L.D., Finn, S.G.: M/M/1 queues with working vacations (M/M/1/WV). Performance Evaluation **50**(1), 41–52 (2002)
13. Tang, Y., Yu, M., Yun, X., Huang, S.: Reliability indices of discrete-time Geox/G/1 queueing system with unreliable service station and multiple adaptive delayed vacations. Journal of Systems Science and Complexity **25**(6), 1122–1135 (2012)
14. Vishnevsky, V.M., Dudin, A.N., Semenova, O.V., Klimenok, V.I.: Performance analysis of the BMAP/G/1 queue with gated servicing and adaptive vacations. Performance Evaluation **68**(5), 446–462 (2011)

Multiserver Queues with Finite Capacity and Setup Time

Tuan Phung-Duc[✉]

Department of Mathematical and Computing Sciences,
Tokyo Institute of Technology, Ookayama, Meguro-ku, Tokyo, Japan
tuan@is.titech.ac.jp

Abstract. Multiserver queues with setup time have been extensively studied because they have application in modelling of power-saving data centers. Although the infinite buffer models are extensively investigated, less attention has been paid to finite buffer models. This paper considers an $M/M/c/K$ queue with setup time for which we suggest a simple and numerically stable recursion for the stationary distribution of the system state. Numerical experiments show various insights into the performance of the system such as performance-energy tradeoff as well as the effect of the capacity on the blocking probability and the mean queue length.

Keywords: Multiserver queue · Setup time · Finite capacity

1 Introduction

The core part of cloud computing is data center where a huge number of servers are available. These servers consume a large amount of energy. Thus, the key issue for the management of these server farms is to minimize the power consumption while keeping acceptable service level for users. It is reported that under the current technology an idle server still consumes about 60% of its peak processing jobs [1]. Thus, the only way to save power is to turn off idle servers. However, off servers need some setup time to be active during which they consume energy but cannot process jobs. Thus, there exists a trade-off between power-saving and performance which could be analyzed by queueing models with setup time.

Recently, motivated by applications in data centers, multiserver queues with setup times have been extensively investigated in the literature. In particular, Gandhi et al. [3] extensively analyze multiserver queues with setup times. They obtain some closed form approximations for the ON-OFF policy where any number of servers can be in the setup mode at a time. As is pointed out in Gandhi et al. [3], from an analytical point of view the most challenging model is the ON-OFF policy where the number of servers in setup mode is not limited. Gandhi et al. [4,5] analyze the $M/M/c$/Setup model with ON-OFF policy using a recursive renewal reward approach. Phung-Duc [11] obtains exact solutions for the

© Springer International Publishing Switzerland 2015
M. Gribaudo et al. (Eds.): ASMTA 2015, LNCS 9081, pp. 173–187, 2015.
DOI: 10.1007/978-3-319-18579-8_13

same model via generating functions and via matrix analytic methods. Slegers et al. [6] propose a heuristic method to decide the timing for the servers to be powered up or down.

Although, the infinite model has been investigated [4,5,11], results for systems with a large number (several hundreds) of servers are not obtained. This motivates us to develop models for large-scale server farms. Furthermore, less attention has been paid on finite buffer multiserver queue with setup time. It should be noted that the results for the latter could be used for the former by letting the capacity tend to infinity. The main aim of our current paper is to present a simple recursion for the stationary distribution of the $M/M/c/K$/Setup model which is more realistic for data centers which typically have a finite buffer. The computational complexity of the scheme is significantly reduced in comparison with that of direct methods. As a result, models with several hundreds of servers are easily analyzed. This allows us to explore new insight into the performance of large scale systems that has not been observed in literature. Recently, we become aware of a closely related paper [2], where the authors suggest a recursive scheme for finite buffer model with threshold control. However, the stability of the numerical scheme is not discussed. In contrast to [2], we suggest here a new recursive scheme whose numerical stability is rigorously proved.

The rest of this paper is organized as follows. Section 2 presents the model in details while Section 3 is devoted to derivation of a recursion for the joint stationary distribution. Section 4 presents some numerical examples showing insights into the performance of the system. Concluding remarks are presented in Section 5.

2   Model

We consider a queueing system with c servers and a capacity of K, i.e., the maximum of K customers can be accommodated in the system. Jobs arrive at the system according to a Poisson process with rate λ. In this system, a server is turned off immediately if it has no job to do. Upon arrival of a job, an OFF server is turned on if any and the job is placed in the buffer. However, a server needs some setup time to be active so as to serve waiting jobs. We assume that the setup time follows an exponential distribution with mean $1/\alpha$. Let j denotes the number of customers in the system and i denotes the number of active servers. The number of servers in setup process is $\min(j - i, c - i)$. Under these assumptions, the number of active servers is smaller than or equal to the number of jobs in the system. Therefore, in this model a server is in either BUSY or OFF or SETUP. We assume that the service time of jobs follows an exponential distribution with mean $1/\mu$. We assume that waiting jobs are served according to a first-come-first-served (FCFS) manner. We call this model an $M/M/c/K$/Setup queue.

The exponential assumptions for the inter-arrival, setup time and service time allow to construct a Markov chain whose stationary distribution is recursively obtainable. It should be noted that we can easily construct a Markov chain for a more general model with MAP arrival and phase-type service and setup time distributions. However, the state space of the resulted Markov chain explodes and thus the analysis is complex.

3   Analysis

In this section, we present a recursive scheme to calculate the joint stationary distribution. Let $C(t)$ and $N(t)$ denote the number of active servers and the number of customers in the system, respectively. It is easy to see that $\{X(t) = (C(t), N(t)); t \geq 0\}$ forms a Markov chain on the state space:

$$\mathcal{S} = \{(i,j); 0 \leq i \leq c, j = i, i+1, \ldots, K-1, K\}.$$

See Figure 1 for transition among states for the case $c = 2$ and $K = 5$.

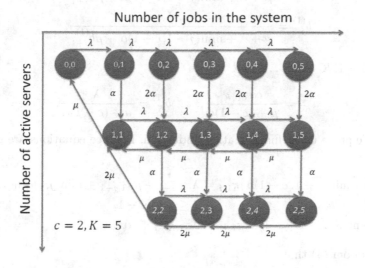

Fig. 1. Transition among states ($c = 2, K = 5$)

Let $\pi_{i,j} = \lim_{t\to\infty} P(C(t) = i, N(t) = j)$ $((i,j) \in \mathcal{S})$ denote the joint stationary distribution of $\{X(t)\}$. In this section, we derive a recursion for calculating the joint stationary distribution $\pi_{i,j}$ $((i,j) \in \mathcal{S})$. The balance equations for states with $i = 0$ read as follows.

$$\lambda \pi_{0,0} = \mu \pi_{1,1},$$
$$(\lambda + \min(j,c)\alpha)\pi_{0,j} = \lambda \pi_{0,j-1}, \qquad j = 1, 2, \ldots, K-1,$$
$$c\alpha \pi_{0,K} = \lambda \pi_{0,K-1}.$$

leading to $\pi_{0,j} = b_j^{(0)} \pi_{0,j-1}$ where $b_j^{(0)} = \lambda/(\lambda + \min(j,c)\alpha)$ $(j = 1, 2, \ldots, K-1)$ and $b_K^{(0)} = \lambda/(c\alpha)$. Furthermore, it should be noted that $\pi_{1,1}$ is calculated using the local balance equation in and out the set $\{(0,j); j = 0, 1, \ldots, K\}$ as follows.

$$\mu\pi_{1,1} = \sum_{j=1}^{K} \min(j,c)\alpha\pi_{0,j}.$$

Remark 1. We have expressed $\pi_{0,j}$ $(j = 1,2,\ldots,K)$ and $\pi_{1,1}$ in terms of $\pi_{0,0}$.

Next, we consider the case $i = 1$.

Lemma 1. *We have*

$$\pi_{1,j} = a_j^{(1)} + b_j^{(1)}\pi_{1,j-1}, \qquad j = 2,3,\ldots,K-1,K,$$

where

$$a_j^{(1)} = \frac{\mu a_{j+1}^{(1)} + \min(j,c)\alpha\pi_{0,j}}{\mu + \lambda + \min(j-1,c-1)\alpha - \mu b_{j+1}^{(1)}}, \qquad (1)$$

$$b_j^{(1)} = \frac{\lambda}{\mu + \lambda + \min(j-1,c-1)\alpha - \mu b_{j+1}^{(1)}}, \qquad (2)$$

for $j = K-1, K-2, \ldots, 2$ and

$$a_K^{(1)} = \frac{c\alpha\pi_{0,K}}{\mu + (c-1)\alpha}, \qquad b_K^{(1)} = \frac{\lambda}{\mu + (c-1)\alpha}.$$

Proof. We prove using mathematical induction. Balance equations are given as follows.

$$(\lambda + \mu + \min(j-1,c-1)\alpha)\pi_{1,j} = \lambda\pi_{1,j-1} + \mu\pi_{1,j+1} + \min(j,c)\alpha\pi_{0,j}, \qquad (3)$$
$$2 \le j \le K-1,$$
$$(\mu + \min(K-1,c-1)\alpha)\pi_{1,K} = \lambda\pi_{1,K-1} + c\alpha\pi_{0,K}. \qquad (4)$$

It follows from (4) that

$$\pi_{1,K} = a_K^{(1)} + b_K^{(1)}\pi_{1,K-1},$$

leading to the fact that Lemma 1 is true for $j = K$. Assuming that Lemma 1 is true for $j+1$, i.e., $\pi_{1,j+1} = a_{j+1}^{(1)} + b_{j+1}^{(1)}\pi_{1,j}$. It then follows from (3) that Lemma 1 is also true for j, i.e., $\pi_{1,j} = a_j^{(1)} + b_j^{(1)}\pi_{1,j-1}$.

Theorem 2. *We have the following bound.*

$$a_j^{(1)} \ge 0, \qquad 0 \le b_j^{(1)} \le \frac{\lambda}{\mu + \min(j-1,c-1)\alpha},$$

for $j = 2,3,\ldots,K-1,K$.

Proof. We use mathematical induction. It is easy to see that the theorem is true for $j = K$. Assuming that the theorem is true for $j+1$, i.e.,

$$a_{j+1}^{(1)} \geq 0, \qquad 0 \leq b_{j+1}^{(1)} \leq \frac{\lambda}{\mu + \min(j, c-1)\alpha}, \qquad j = 1, 2, \ldots, K-1.$$

Thus, we have $\mu b_{j+1}^{(1)} < \lambda$. From this inequality, (1) and (2), we obtain

$$b_j^{(1)} \leq \frac{\lambda}{\mu + \min(j-1, c-1)\alpha},$$

and $a_j^{(1)} \geq 0$.

It should be noted that $\pi_{2,2}$ can be calculated using the local balance between the flows in and out the set of states $\{(i,j); i = 0, 1, j = i, i+1, \ldots, K\}$ as follows.

$$2\mu\pi_{2,2} = \sum_{j=2}^{K} \min(j-1, c-1)\alpha\pi_{1,j}.$$

Remark 2. We have expressed $\pi_{1,j}$ $(j = 1, 2 \ldots, K)$ and $\pi_{2,2}$ in terms of $\pi_{0,0}$.

We consider the general case where $2 \leq i \leq c-1$. Similar to the case $i = 1$, we can prove the following result by mathematical induction.

Lemma 3. *We have*

$$\pi_{i,j} = a_j^{(i)} + b_j^{(i)}\pi_{i,j-1}, \qquad j = i+1, i+2, \ldots, K-1, K,$$

where

$$a_j^{(i)} = \frac{i\mu a_{j+1}^{(i)} + \min(c-i+1, j-i+1)\alpha\pi_{i-1,j}}{\lambda + \min(c-i, j-i)\alpha + i\mu - i\mu b_{j+1}^{(i)}}, \qquad (5)$$

$$b_j^{(i)} = \frac{\lambda}{\lambda + \min(c-i, j-i)\alpha + i\mu - i\mu b_{j+1}^{(i)}}, \qquad (6)$$

and

$$a_K^{(i)} = \frac{(c-i+1)\alpha\pi_{i-1,K}}{(c-i)\alpha + i\mu}, \qquad b_K^{(i)} = \frac{\lambda}{(c-i)\alpha + i\mu}.$$

Proof. The balance equation for state (i, K) is given as follows.

$$((c-i)\alpha + i\mu)\pi_{i,K} = \lambda\pi_{i,K-1} + (c-i+1)\alpha\pi_{i-1,K},$$

leading to the fact that Lemma 3 is true for $j = K$. Assuming that

$$\pi_{i,j+1} = a_{j+1}^{(i)} + b_{j+1}^{(i)}\pi_{i,j}, \qquad j = i+1, i+2, \ldots, K-1.$$

It then follows from

$$(\lambda + \min(c-i, j-i)\alpha + i\mu)\pi_{i,j}$$
$$= \lambda\pi_{i,j-1} + i\mu\pi_{i,j+1} + \min(c-i+1, j-i+1)\alpha\pi_{i-1,j},$$
$$j = K-1, K-2, \ldots, i+1,$$

that

$$\pi_{i,j} = a_j^{(i)} + b_j^{(i)}\pi_{i,j-1}.$$

Theorem 4. *We have the following bound.*

$$a_j^{(i)} > 0, \qquad 0 < b_j^{(i)} < \frac{\lambda}{i\mu + \min(j - i, c - i)\alpha},$$

for $j = i + 1, i + 2, \ldots, K - 1, i = 1, 2, \ldots, c - 1.$

Proof. We also prove using mathematical induction. It is clear that Theorem 4 is true for $j = K$. Assuming that Theorem 4 is true for $j + 1$, i.e.,

$$a_{j+1}^{(i)} > 0, \qquad 0 < b_{j+1}^{(i)} < \frac{\lambda}{i\mu + \min(j + 1 - i, c - i)\alpha},$$

for $j = i + 1, i + 2, \ldots, K - 1, i = 1, 2, \ldots, c - 1.$ It follows from the second inequality that $i\mu b_{j+1}^{(i)} < \lambda$. This together with formulae (5) and (6) yield the desired result.

It should be noted that $\pi_{i+1,i+1}$ is calculated using the following local balance equation in and out the set of states:

$$\{(k, j); k = 0, 1, \ldots, i; j = k, k + 1, \ldots, K\}$$

as follows.

$$(i + 1)\mu\pi_{i+1,i+1} = \sum_{j=i+1}^{K} \min(j - i, c - i)\alpha\pi_{i,j}.$$

Remark 3. We have expressed $\pi_{i,j}$ $(i = 0, 1, \ldots, c - 1, j = i, i + 1, \ldots, K)$ and $\pi_{i+1,i+1}$ in terms of $\pi_{0,0}$.

Finally, we consider the case $i = c$. Balance equation for state (c, K) yields,

Lemma 5. *We have*

$$\pi_{c,j} = a_j^{(c)} + b_j^{(c)}\pi_{c,j-1}, \qquad j = c + 1, c + 2, \ldots, K - 1,$$

where

$$a_j^{(c)} = \frac{c\mu a_{j+1}^{(c)} + \alpha\pi_{c-1,j}}{\lambda + c\mu - c\mu b_{j+1}^{(c)}}, \qquad j = K - 1, K - 2, \ldots, c + 1, \qquad (7)$$

$$b_j^{(c)} = \frac{\lambda}{\lambda + c\mu - c\mu b_{j+1}^{(c)}}, \qquad j = K - 1, K - 2, \ldots, c + 1, \qquad (8)$$

and

$$a_K^{(c)} = \frac{\alpha\pi_{c-1,K}}{c\mu}, \qquad b_K^{(c)} = \frac{\lambda}{c\mu}.$$

Proof. The global balance equation at state (c, K) is given by

$$c\mu\pi_{c,K} = \alpha\pi_{c-1,K} + \lambda\pi_{c,K-1},$$

leading to

$$\pi_{c,K} = a_K^{(c)} + b_K^{(c)}\pi_{c,K-1}.$$

Assuming that $\pi_{c,j+1} = a_{j+1}^{(c)} + b_{j+1}^{(c)}\pi_{c,j}$, it follows from the global balance equation at state (c,j),

$$(\lambda + c\mu)\pi_{c,j} = \lambda\pi_{c,j-1} + c\mu\pi_{c,j+1} + \alpha\pi_{c-1,j}, \qquad j = c+1, c+2, \dots, K-1,$$

that $\pi_{c,j} = a_j^{(c)} + b_j^{(c)}\pi_{c,j-1}$ for $j = c+1, c+2, \dots, K$.

Theorem 6. *We have the following bound.*

$$a_j^{(c)} > 0, \qquad 0 < b_j^{(c)} < \frac{\lambda}{c\mu}, \qquad j = c+1, c+2, \dots, K-1.$$

Proof. We also prove using mathematical induction. It is clear that Theorem 6 is true for $j = K$. Assuming that Theorem 6 is true for $j+1$, i.e.,

$$a_{j+1}^{(c)} > 0, \qquad 0 < b_{j+1}^{(c)} < \frac{\lambda}{c\mu}, \qquad j = c+1, c+2, \dots, K-1.$$

It follows from the second inequality that $c\mu b_{j+1}^{(c)} < \lambda$. This together with formulae (7) and (8) yield the desired result.

We have expressed all the probability $\pi_{i,j}$ ($(i,j) \in \mathcal{S}$) in terms of $\pi_{0,0}$ which is uniquely determined by the normalizing condition.

$$\sum_{(i,j)\in\mathcal{S}} \pi_{i,j} = 1.$$

Remark 4. We see that the computational complexity order for $\{\pi_{i,j}; (i,j) \in \mathcal{S}\}$ is $O(cK)$. A direct method for solving the set of balance equations requires the complexity of $O(c^3 K^3)$ while a level-dependent QBD approach (See Phung-Duc et al. [8]) needs the computational complexity of $O(Kc^3)$. We also observe that the recursion scheme of this paper is numerically stable since it manipulates only positive numbers (See Theorems 2, 4 and 6).

4 Performance Measures and Numerical Examples

4.1 Performance Measures

Let P_B denote the blocking probability. We have

$$P_B = \sum_{i=0}^{c} \pi_{i,K}.$$

Let π_i denote the stationary probability that there are i active servers, i.e., $\pi_i = \sum_{j=i}^{K} \pi_{i,j}$. Let $\mathbb{E}[A]$ and $\mathbb{E}[S]$ denote the mean number of active servers and that in setup mode, respectively. We have

$$\mathbb{E}[A] = \sum_{i=1}^{c} i\pi_i, \qquad \mathbb{E}[S] = \sum_{i=0}^{c} \sum_{j=i}^{K} \min(j-i, c-i)\pi_{i,j}.$$

The power consumption per a unit time for the model with setup time is given by

$$Cost_{on-off} = C_a \mathbb{E}[A] + C_s \mathbb{E}[S], \tag{9}$$

where C_a and C_s are the cost per a unit time for an active server and a server in setup mode, respectively.

For comparison, we also find the power consumption per a unit time for the corresponding ON-IDLE model, i.e., $M/M/c/K$ without setup times. Letting p_i $(i = 0, 1, \ldots, K-1, K)$ denote the stationary probability that there are i customers in the system, we have

$$p_i = \left(\frac{\lambda}{\mu}\right)^i \frac{1}{i!} p_0, \quad i = 0, 1, \ldots, c,$$

$$p_i = p_c \left(\frac{\lambda}{c\mu}\right)^{i-c}, \quad i = c, c+1, \ldots, K-1, K,$$

where p_0 is determined by the normalization condition $\sum_{i=0}^{K} p_i = 1$. Let $\mathbb{E}[\widehat{A}]$ denote the mean number of active servers, we have

$$\mathbb{E}[\widehat{A}] = \sum_{i=0}^{K} \min(i, c) p_i = \frac{\lambda(1-p_K)}{\mu},$$

where the second equality is due to Little's law. Therefore, the mean number of idle servers is given by $c - \mathbb{E}[\widehat{A}]$. Thus, for this model, the power consumption per a unit time is given by

$$Cost_{on-idle} = C_a \mathbb{E}[\widehat{A}] + (c - \mathbb{E}[\widehat{A}])C_i. \tag{10}$$

where C_i is the cost per a unit time for an idle server.

Let $\mathbb{E}[N]$ denote the mean number of customers in the system. We have

$$\mathbb{E}[N] = \sum_{i=0}^{c} \sum_{j=i}^{K} \pi_{i,j} \times j.$$

Let $\mathbb{E}[T]$ denote the mean response time of a customer. We have

$$\mathbb{E}[T] = \frac{\mathbb{E}[N]}{\lambda(1-P_B)}.$$

4.2 M/M/c/c System

We consider the following parameter setting: $c = K$, $\mu = \alpha = 1$. Furthermore, we set the cost for an active server and that for a setup server as $C_a = C_s = 1$ as in [7]. The cost for an idle server is $C_i = 0.6$ because an idle server still consumes 60% energy of its peak processing a job [1]. We investigate the power consumption for the M/M/c/K/Setup queue and its corresponding M/M/c/K model by (9) and (10), respectively. Figures 2 and 4 represent the blocking probability and power consumption against $\rho = \lambda/(c\mu)$ for the case $c = K = 50$ while Figures 3 and 5 represent those for the case $c = K = 500$. We observe that the blocking probability P_B decreases with α and is bounded from below by that of the corresponding ON-IDLE model (p_K). We also observe that our numerical scheme is stable since it can calculate the blocking probability of order 10^{-17}.

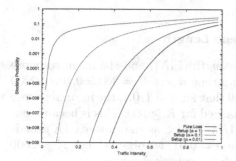

Fig. 2. Blocking probability against ρ $(c = 50)$

Fig. 3. Blocking probability against ρ $(c = 500)$

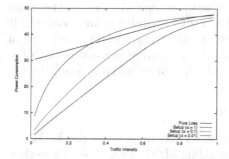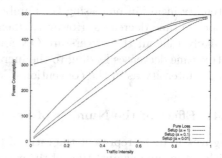

Fig. 4. Power consumption against ρ $(c = 50)$

Fig. 5. Power consumption against ρ $(c = 500)$

We observe from Figures 4 and 5 that the power consumption increases with the traffic intensity ρ as expected. Furthermore, for the case $\alpha = 1, 0.1$, the ON-OFF policy outperforms the ON-IDLE one for any value of ρ. As for the case

$\alpha = 0.01$ there exist a range in which the power consumption of the ON-IDLE model is smaller than that of the ON-OFF model. Furthermore, the range for $c = 50$ is larger than that of $c = 500$. This suggests that the ON-OFF policy is more advanced in large-scale systems.

Figures 6 and 7 represent the mean number of setup servers $\mathbb{E}[S]$ against traffic intensity for the case $c = 50$ and $c = 500$, respectively. We observe that there exists some $\widehat{\rho}_\alpha$ such that $\mathbb{E}[S]$ increases with ρ in the range $(0, \widehat{\rho}_\alpha)$ while $\mathbb{E}[S]$ decreases with ρ for the range $(\widehat{\rho}_\alpha, 1)$. This is because when the traffic intensity is small, many servers are turned off. As a result, increasing the traffic intensity (number of arriving customers) incurs in the increase in the mean number of servers in setup. However, when the traffic intensity is large enough, almost the servers are likely on for all the time. Thus, the effect of setup is less and then the mean number of servers in setup time decreases with the traffic intensity.

4.3   Mean Response Time and Queue Length

In this section, we show the mean queue length ($\mathbb{E}[N]$) and the mean response time ($\mathbb{E}[T]$) of the M/M/100/K with setup time where $K = 200, 500, 1000, 2000$ and 3000. We observe from Figures 8 and 9 that for $\alpha = 1, 0.1$, the mean response time and the mean queue length are unchanged for $K \geq 500$. This is because our system converges to the corresponding M/M/100/∞ as the capacity (K) tends to infinity. However, for the curves where $\alpha = 0.01$, we observe that $K = 2000$ is not large enough to approximate the infinite capacity system.

We observe in all the curves that the mean queue length increases with the traffic intensity. On the other hand, the mean response time decreases with ρ when ρ is small while it increases with ρ when ρ is large. This is because at low traffic intensity, the effect of setup time is large. Thus, increasing the traffic intensity incurs in increasing the number of setup servers. As a result the mean response time decreases. However, when the traffic intensity is large enough, it is likely that all the servers are ON for all the time. As a result, the effect of setup time decreases leading to the increase of the mean response time with the traffic intensity as in the conventional M/M/c/K system without setup time.

4.4   Effect of the Number of Servers

Figures 10 to 13 represent the ratio of the power consumption of the M/M/c/c with setup time against that of the corresponding M/M/c/c without setup time ($Cost_{on-off}/Cost_{on-idle}$) for $\rho = 0.3, 0.5, 0.7$ and 0.9. We observe that under all considered traffic intensities, the ratio is less than one for $\alpha = 1, 0.1$ meaning that the former is less power-consuming than the latter for $\alpha = 1$ and 0.1. On the other hand, for $\alpha = 0.01$, the latter outperforms the former for a wide range of c. This may be due to the fact that a large portion of customers are lost due to the slow setup ($1/\alpha = 100$). We observe in the case $\rho = 0.3, 0.5$ and 0.7 that the power consumption ratio decreases with c.

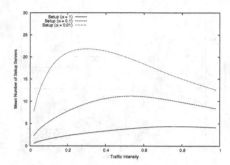

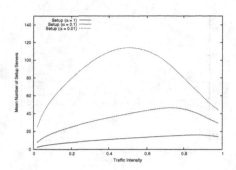

Fig. 6. Mean number of setup servers against ρ $(c = 50)$

Fig. 7. Mean number of setup servers against ρ $(c = 500)$

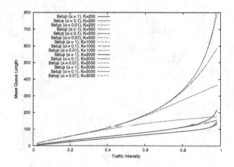

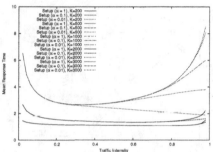

Fig. 8. Mean queue length against ρ $(c = 100)$

Fig. 9. Mean response time against ρ $(c = 100)$

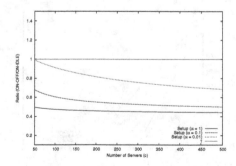

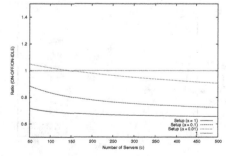

Fig. 10. Ratio of power consumption $(\rho = 0.3)$

Fig. 11. Ratio of power consumption $(\rho = 0.5)$

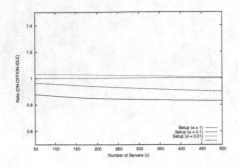

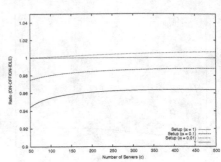

Fig. 12. Ratio of power consumption ($\rho = 0.7$)

Fig. 13. Ratio of power consumption ($\rho = 0.9$)

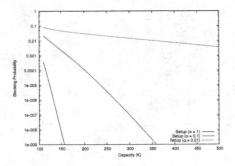

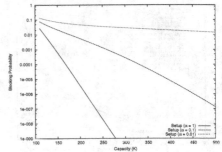

Fig. 14. Blocking probability against K ($\rho = 0.7, c = 100$)

Fig. 15. Blocking probability against K ($\rho = 0.9, c = 100$)

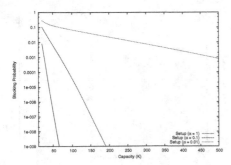

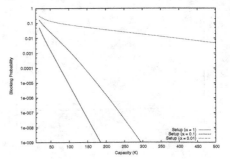

Fig. 16. Blocking probability against K ($\rho = 0.7, c = 10$)

Fig. 17. Blocking probability against K ($\rho = 0.9, c = 10$)

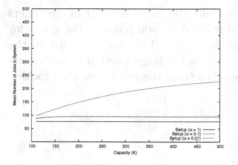

Fig. 18. Mean number of jobs in system against K ($\rho = 0.7, c = 100$)

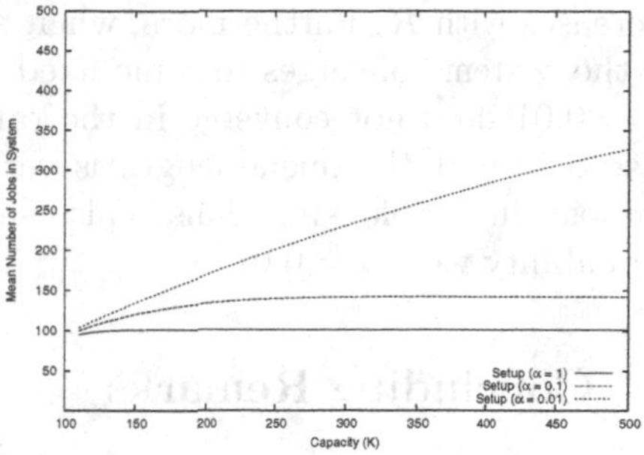

Fig. 19. Mean number of jobs in system against K ($\rho = 0.9, c = 100$)

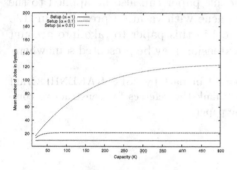 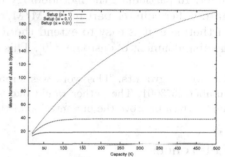

Fig. 20. Mean number of jobs in system against K ($\rho = 0.7, c = 10$)

Fig. 21. Mean number of jobs in system against K ($\rho = 0.9, c = 10$)

4.5   Effect of the Capacity

In this section, we show the influence of the capacity K on the performance of the system. We consider the cases where $\rho = 0.7$ and $\rho = 0.9$ while $c = 10$ and 100. Figures 14 to 17 represent the blocking probability against K for the $c = 100, 10$ and $\rho = 0.7, 0.9$. We observe in all these graphs that the blocking probability geometrically decreases in K. We observe in the curves for $\alpha = 1, 0.1$ that the blocking probability is sensitive to K in the sense that it decreases with K at a high speed. On the other hand, we observe that the blocking blocking probability for the case $\alpha = 0.01$ is less sensitive to K in comparison with the cases $\alpha = 1, 0.1$.

Figures 18 to 21 represent the mean number of customers in the system against K for the $c = 100, 10$ and $\rho = 0.7, 0.9$. We observe in the graphs for $\rho = 0.7$ that the mean number of customers in the system increases with K and then converges to some fixed value. This is intuitive because our system converges to the M/M/c/∞ with setup time when $K \rightarrow \infty$. In the graphs for $\rho = 0.9$ we also observe that the mean number of customers in the system increases with K. Furthermore, when $\alpha = 1, 0.1$ the mean number of customers in the system converges to some fixed value for $K < 500$ however the curve for $\alpha = 0.01$ does not converge in the range $K < 500$. This suggests that in the case $\alpha = 0.01$ the queue length is very long and a large portion of customers are lost due to blocking. This is also supported from the curves for the blocking probability with $\alpha = 0.01$.

5   Concluding Remarks

We present a simple recursion to calculate the stationary distribution of the system state of an $M/M/c/K$ queue with setup time for data centers. The computational complexity order of the algorithm is only $O(cK)$. The methodology of this paper can be applied for various variant models with setup time and finite buffer. In particular, the methodology of this paper can also be applied to the finite buffer counter part of the $M/M/c$ queue with vacation presented in [12]. Furthermore, it is easy to extend the model in this paper to take into account the abandonment of customers [9]. This extension may be presented somewhere.

Acknowledgments. This work was supported in part by JSPS KAKENHI Grant Number 2673001. The author would like to thank the referees for constructive comments which improve the presentation of the paper.

References

1. Barroso, L.A., Holzle, U.: The case for energy-proportional computing. Computer **40**(12), 33–37 (2007)
2. Kuehn, P.J., Mashaly, M.E.: Automatic energy efficiency management of data center resources by load-dependent server activation and sleep modes. Ad Hoc Networks **25**, 497–504 (2015)
3. Gandhi, A., Harchol-Balter, M., Adan, I.: Server farms with setup costs. Performance Evaluation **67**, 1123–1138 (2010)
4. Gandhi, A., Doroudi, S., Harchol-Balter, M., Scheller-Wolf, A.: Exact analysis of the M/M/k/setup class of Markov chains via recursive renewal reward. In: Proceedings of the ACM SIGMETRICS, pp. 153–166. ACM (2013)
5. Gandhi, A., Doroudi, S., Harchol-Balter, M., Scheller-Wolf, A.: Exact analysis of the M/M/k/setup class of Markov chains via recursive renewal reward. Queueing Systems **77**(2), 177–209 (2014)
6. Slegers, J., Thomas, N., Mitrani, I.: Dynamic server allocation for power and performance. In: Kounev, S., Gorton, I., Sachs, K. (eds.) SIPEW 2008. LNCS, vol. 5119, pp. 247–261. Springer, Heidelberg (2008)
7. Mitrani, I.: Managing performance and power consumption in a server farm. Annals of Operations Research **202**(1), 121–134 (2013)
8. Phung-Duc, T., Masuyama, H., Kasahara, S., Takahashi, Y.: A simple algorithm for the rate matrices of level-dependent QBD processes. In: Proceedings of the 5th International Conference on Queueing Theory and Network Applications (QTNA2010), Beijing, China, pp. 46–52. ACM, New York (2010)

9. Phung-Duc, T.: Impatient customers in power-saving data centers. In: Sericola, B., Telek, M., Horváth, G. (eds.) ASMTA 2014. LNCS, vol. 8499, pp. 185–199. Springer, Heidelberg (2014)
10. Phung-Duc, T.: Server farms with batch arrival and staggered setup. In: Proceedings of the Fifth Symposium on Information and Communication Technology, pp. 240–247. ACM (2014)
11. Phung-Duc, T.: Exact solution for M/M/c/Setup queue (2014). http://arxiv.org/abs/1406.3084
12. Tian, N., Li, Q.L., Gao, J.: Conditional stochastic decompositions in the M/M/c queue with server vacations. Stochastic Models 15, 367–377 (1999)

Power Consumption Analysis of Replicated Virtual Applications

Pietro Piazzolla[1]([✉]), Gianfranco Ciardo[2], and Andrew Miner[2]

[1] Department of Electronics, Information and Bioengineering,
Politecnico di Milano, Milano, Italy
`pietro.piazzolla@polimi.it`
[2] Department of Computer Science, Iowa State University, Ames, USA
`{ciardo,asminer}@iastate.edu`

Abstract. The search for green IT has inspired a wide spectrum of techniques for power management. In a data center where computational power is provided by means of virtualised resources, like virtual machines, the policy to allocate them on physical servers can strongly impact the power consumption of the entire system. We propose a generalised stochastic Petri net model to investigate the contribution to energy efficiency due to different allocation and deallocation policies.

Keywords: Energy efficiency · Generalised stochastic Petri nets · Virtualised datacenters · Allocation policies · Performance evaluation

1 Introduction

Research to improve IT infrastructure sustainability has inspired a wide spectrum of techniques for power management that exploit, albeit in different ways, two types of basic mechanisms: the dynamic scaling of system components' performance (Dynamic Performance Scaling) and the dynamic hibernation of components (Dynamic Component Deactivation), see e.g.[1]. The main assumption underlying these techniques is that a system experiences a workload that varies over time, allowing for component adjustments. Among the techniques developed to improve energy efficiency in data centers, there are some that work on the system load, providing algorithms, heuristics or policies to schedule it among several servers. In a data center where computational power is provided to users by means of virtualised resources, for example virtual machines (VM), the policy followed to allocate them on physical hosts, or machines (PM), can strongly impact the power consumption of the whole system. In particular, by using different VM placement strategies, it is possible to control host utilization. For example, consolidating several VMs on a single PM allows some other physical machine to be in an idle state, thus lowering system energy consumption. In several works [3], the instantaneous power consumption is shown to have a linear

© Springer International Publishing Switzerland 2015
M. Gribaudo et al. (Eds.): ASMTA 2015, LNCS 9081, pp. 188–202, 2015.
DOI: 10.1007/978-3-319-18579-8_14

relation with CPU utilization, but even if the utilization is zero, the machine still consumes a large amount of energy just for being on and ready to process incoming requests. The power required by each idle server, P_{idle}, can heavily affect the consumption of the entire system. Strategies like consolidation can be used to reduce P_{idle} by sharing its contribution among all the VMs running on the same server. However, consolidation alone will only increase the number of idle machines: to be effective, it must be used in conjunction with a Dynamic Component Deactivation strategy, in order not to pay the power consumption P_{idle} for those hosts that are not utilised.

In this paper we are interested in studying how different VM allocation and release strategies impact energy consumption of a virtualised datacenter. Placement scheduling strategies, or *policies*, define from which available host to allocate the resources for a newly requested VM, while deallocation policies define from which server to release no longer required ones. We focus on systems running groups of identical VM instances, all replicating the services of the same application. Such an application is able to scale its number of instances according to the workload it handles. Since all the VMs are identical replicas and their allocation and deallocation depends only on the number of requests the application receives, it is possible to develop different strategies with different energy efficiency outcomes.

We propose a generalised stochastic Petri net (GSPN) model of a virtualised datacenter to measure the capability of each policy to reduce the number of PMs powered on, thus reducing P_{idle} consumption. This paper is one of the few to focus on the reduction of this power index, as well as one of the few to explicitly include deallocation policies in its analysis .

The rest of this paper is organized as follows. Section 2 gives a brief overview of related literature. Section 3 presents the main assumptions about the specific problem addressed in this paper and the analyzed policies. Section 4 presents the model of the virtualised datacenter in terms of its parameters and discusses the power consumption measures we adopt. The Petri net model is the focus of Section 5, while Section 6 presents results. Section 7 concludes the paper.

2   Related Work

A common problem in data center management is resource allocation in the presence of workloads having fluctuating intensities. In the literature, there are several works that deal with the optimal allocation of resources in virtual environments, aiming at different goals and exploiting different techniques. Of these works, many exploit probabilistic techniques and models [2,14] to maximise selected performance indices.

Motivated by the need to understand and reduce energy waste and its related costs, in the last decade, researchers focused on devising techniques to optimise the power management of servers in large data centers. An exhaustive survey on this specific topic can be found in [1] and references therein. Early works on power consumption [11] propose policies to dynamically turn on or off cluster nodes, according to the system workload, but without addressing the issues introduced by virtualization. Starting from [10], power management techniques have been explored in the context of virtualised systems. The problem of dynamic provisioning of VMs for multitier web applications according to the current workload (number of incoming requests), in an efficient resources management perspective, is the topic of [8]. In several works [12], average data center power consumption is optimised by means of diverse power management policies. Often the proposed solutions exploits consolidation and migration of VMs as opposed to this paper. In [13] migration is also not considered, as the objective is to improve the utilization of resources, leading to reduced energy consumption. Each application can be deployed using several VMs instantiated on different PMs. The resources are allocated to applications proportionally according to the applications' priorities. A similar approach, studying power consumption and replication of services, is presented in [9] and extended in [4].

3   Scenario

The computational and storage power of a datacenter is usually leveraged to execute *applications* whose services can be accessed by end users. We focus on those applications able to automatically scale their demand of system resources to satisfy a different workload of requests. If the resources of the system are provided in form of VMs, and the workload increases, then further VMs are allocated to that particular application. We consider that the VMs providing the application services can be replicated multiple times to serve a larger number of requests. These replicas are identical to one another, providing the same services and are used to scale the system. For the purpose of this work, the life time of a VM, i.e., the interval of time between its instantiation and its deallocation, is orders of magnitude higher than that of the single incoming job it serves. Once the application's request for further instances is acknowledged by the system, the new VM(s) instance(s) must be assigned to the available PM(s).

Different scheduling policies can be implemented to allocate the resources for new VMs among the available servers of the datacenter [15]. Some of the most common are:

- *Random*: a randomly determined PM is selected to host the new VM.
- *Round-robin*: the VM is placed on the next available PM in a sequence.
- *Least loaded server*: the VM is placed on the PM with the most available resources.
- *Most loaded server*: the VM is placed on the PM with the fewest available resources.

In the literature the allocation of VMs among the PMs is performed using optimization algorithms, but these solutions often require time and knowledge about the load of each application, a knowledge that is rarely available beforehand.

When the peak of requests terminates, the application releases all the excess VMs, scaling down its resources requirements. Different release policies can be

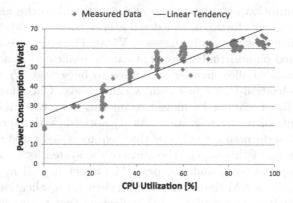

Fig. 1. Power consumption vs Utilization [3]

implemented to determine which VM is to be turned off. In this work we consider the following release policies: a randomly determined PM is selected to release the VM (*Random*), the VM is removed from the PM with the most available resources (*Least loaded server*), the VM is removed from the PM with the fewest available resources (*Most loaded server*).

The specific policy to allocate and release resources for VMs determines the utilization levels of the various hosts in a datacenter. In Fig. 1, from [3], the typical relation between the utilization of a server and its power consumption is shown. Although a less utilised server will consume less energy, as it can be seen, even if the utilization is zero, the machine still consumes energy simply because it is powered on. To reduce their P_{idle} contribution, unutilised servers must be put in an state in which their power consumption is negligible or null, using one or more Dynamic Component Deactivation techniques. Policies can be defined to determine when such servers are to be turned on or off. We assume that a server is turned off when it hosts no VMs, and is turned on when no resources are available in the pool of current active servers to allocate a new VM.

We want to study how the policies that determine the placement and release of virtual machines among PMs can influence the power consumption of a datacenter. In particular, allocating and deallocating resources for VM instances according to a specific strategy can determine different levels of utilization among the servers that may put a higher or lower number of them in idle state. Since idle PMs can be deactivated to save power, policies can be seen as a tool to increase power efficiency by reducing the P_{idle} contribution.

4 Model Description

Consider a system composed by a number M (large but finite) of homogeneous PMs, each able to allocate C resources for running VMs. These PMs can be either *powered on* or *powered off*. In the first case, they are able to serve requests while in the second case they are not, but will consume (little or) no power. We assume the power consumption of a datacenter is proportional to the number of PMs powered on. We ignore startup or shutdown energy costs. The system is used by K different applications (basically classes of VMs). The services each application provides are based on a virtual server that can be replicated a maximum of N_k ($1 \leq k \leq K$) times to allow heavier workloads to be served. N_k represents the degree of parallelization that application k can achieve. Once instantiated on a server m, a replica n cannot be moved to another server. That is, our model does not account for server migration. As [1] noticed, VM migration leads to time delays and performance overheads that require a dedicated analysis which is outside the scope of this paper. The state of the system is a collection of N_k variables per application k, and one per PM m. Let us call $n_{k,i}$ the variable associated to the i-th VM that is hosting services for application k (with $1 \leq i \leq N_k$). Its domain is $n_{k,i} \in \{0, \ldots, M\}$ indicating that the replica is not used (if $n_{k,i} = 0$) or the number of the PM to which it is allocated ($n_{k,i} = m$). The boolean variable γ_m indicates whether server m is powered off ($\gamma_m = 0$) or not ($\gamma_m = 1$). We assume that there are always resources available in the system to satisfy a request to allocate one more VM, that is:

$$ M \cdot C \geq \sum_{k=1}^{K} N_k. \tag{1} $$

Two different types of events can change the state of the system:

- α events: request to allocate (α_k^+) or deallocate (α_k^-) a new VM. Allocating a new instance n for an application k is always possible unless $n = N_k$. Deallocating is always possible provided there is at least one $n \neq 0$ for an application k.
- β events: requests to power on (β^+) or off (β^-) a PM. Powering on a new PM occurs when no resources are available to allocate a new VM. On the converse, when a PM has no VMs it can be shut down. Powered off PMs do not contribute to system power consumption.

For each application k ($1 \leq k \leq K$), the trigger of α_k^+ events follows a Poisson Process with an interarrival rate λ_k. Events α_k^- for application k trigger following a Poisson Process with service rate μ_k. Since we consider the lifetime of a VM to be orders of magnitude longer than that of the requests it serves, an α_k^- event is fired when the number of requests to application k can be served by one fewer VM. The allocation policies listed in Section 3 define how the state of the system changes after the occurrence of an event α or β. According to the focus of this study, the most significant parameter that we want to minimize is the P_{idle} required by a running PM. We define the energy consumption function or *efficiency* $E(t)$ as the number of PMs powered off at time t. In particular, since we assume that no time is required to start or stop a PM and that β events are immediate, then new PMs are instantaneously available while PMs without VMs on it are automatically turned off. Let $n_{VM}(m, t)$ be the number of VMs allocated on PM m at time t, and let us denote $\mathbf{1}(\phi)$ the indicator function that returns 1 if predicate ϕ is true, and 0 otherwise. We can then define $E(t)$ as:

$$E(t) = \sum_{m=1}^{M} \mathbf{1}(n_{VM}(m,t) = 0) \tag{2}$$

To compute the power consumption of the running PMs, instead, a more accurate approach is required. It has been shown [7] that a good approximation of the power consumption of a server can be described by a linear function of the utilization:

$$P(U) = P_{idle} + U \cdot (P_{max} - P_{idle}) \tag{3}$$

where P_{max} is the maximum power consumption that a given PM will have. We can approximate the utilization as $U(n) = n/C$, where n is the number of VMs running on the considered PM. To measure the energy consumption of the system we define an *estimated power consumption function* $P(t)$ that considers the number of PMs powered on at time t and the number of VMs running on them. We can then define $P(t)$ as:

$$P(t) = \sum_{m=1}^{M} \left(\mathbf{1}(n_{VM}(m,t) \geq 1) \cdot P_{idle} + \frac{n_{VM}(m,t)}{C} \cdot (P_{max} - P_{idle}) \right) \tag{4}$$

If the P_{idle} consumption is higher than the power consumption per VM (that is, considering P_{max} and PM capacity C), then Eq. 2 correctly accounts for this expenditure. Otherwise, if the power consumption per VM is higher than P_{idle} consumption, Eq. 4 provides a better approximation because it accounts for different PM types.

5 Petri Net

Fig. 2 shows the proposed GSPN model of the system. Places p_k represent virtual machines not (yet) requested for allocation by application k and may contain a number of tokens up to the maximum number of replicas allowed per application, N_k. Places p'_k are their complementary places ($\#p'_k = N_k - \#p_k$, where $\#x$ signifies the number of tokens in place x) and contain the number of running VMs for each application k.

The available PMs are represented by PM_m places, each holding up to C tokens. These tokens represent the resources each PM_m can provide to run VMs. For each applications k, places $VM_{k,m}$ represent the number of running VMs an application has allocated on PM m. When transition α_k^+ fires for application k, the request for a new VM is issued (token in q_k) and, according to the given policy (see Section 5.1), one of the immediate transitions $g_{k,m}$ fires, placing the VM on one of the available PMs (adding one token in $VM_{k,m}$). When requests arrive, immediate transitions $g_{k,m}$ can fire only if there are resources available in the corresponding PM m (tokens in $PM_m > 0$). When the α_k^- transition is fired for application k, the request for stopping one of its running VMs is issued. This transition may fire only if there are running VMs for the application (tokens in $p'_k > 0$). Once there are tokens in r_k, VMs and PMs resources can be released following a policy that enables one of the immediate transitions $h_{k,m}$.

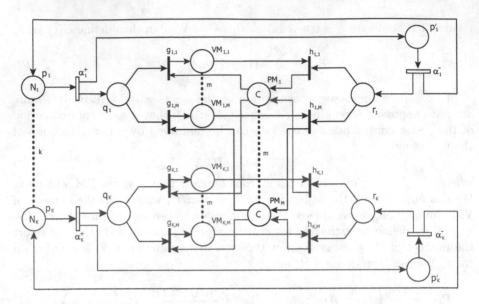

Fig. 2. Petri net model of the system. The dotted lines indicate replications of subnets due to replication of places.

5.1   Modelling the Policies

Random placement and release policies introduced in Section 3 can be implemented with no changes to the model in Fig. 2, as the random selection of one of the available $VM_{k,m}$ places naturally corresponds to PN semantics.

Most-Least Loaded Server placement policies can be implemented as a guard introduced to the immediate transitions $g_{k,m}$, while the corresponding release policies require a guard for $h_{k,m}$ transitions. If we consider as the *least loaded server* PM_m the one with the highest capacity available we can define, for the model in Fig. 2 , the following guards as the *Least loaded server* placement and release policies:

$$g_{k,m} \equiv h_{k,m} \equiv \bigwedge_{i \neq m} (\#PM_m \geq \#PM_i) \tag{5}$$

where the $\wedge$ symbol is the logical *AND* among the propositions.

The *most loaded server* is the PM_m with the lowest capacity available, then for the model in Fig. 2, we can define the following guards as the *Most loaded server* placement and release policies:

$$g_{k,m} \equiv h_{k,m} \equiv \bigwedge_{i \neq m} (\#PM_m \leq \#PM_i) \tag{6}$$

Round Robin Policy. To be implemented as a placement policy requires the model in Fig. 2 to be changed. The model in Fig. 3 takes into account such changes and models the policy with the addition of a subnet whose places and

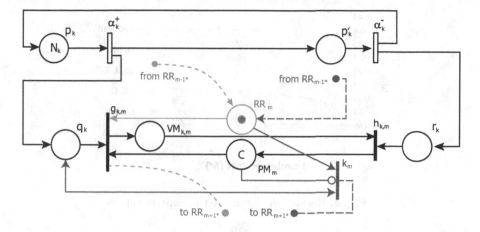

Fig. 3. Petri net of the model with round robin allocation policy

arcs are grey. The token in RR_m place represents the server m next in line to receive the allocation of a new VM. Once server m receives the VM, the round robin token is moved to the next server. The $*$ character indicates that once the token is in RR_M, that is the last place of the round robin cycle, the next PM in line will be PM_1. In case a given PM_m runs out of resources, the subnet depicted in red make possible for the round robin token to skip that m and find the next available PM, thanks to the firing of k_m transitions. Bi-directional arcs (q_k, k_m) enable this skipping only when a new VM allocation is issued, thus avoiding the round robin token to loop forever when there are no free PMs. In this paper, we implement round robin as a placement policy only.

6 Experimental Results

This section presents results obtained solving the models of the previous sections using the SMART [5] tool (Stochastic Model checking Analyzer for Reliability and Timing). SMART takes in input a GSPN and generates the underlying continuous-time Markov chain (CTMC). To compute the steady-state distribution of the CTMC for models in Fig. 2 and 3, we use the Gauss-Seidel option (#Solver GAUSS_SEIDEL) with an exact symbolic representation of the transition rate matrix (#SolutionType EXACT_EVMDD).

We ran tests with different system configurations, each implementing a specific pair of "allocation–deallocation" policies, using the codes Rnd (random), RR (round robin), LL (least-loaded), ML (most-loaded). Some of these combinations correspond to real datacenters configurations, other are considered for comparative purposes. We used an i7 ASUS machine, running Ubuntu 14.04 OS.

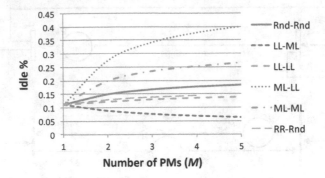

Fig. 4. Results for the first set of experiments

6.1 Two Application Classes

For the first set of experiments, consider a simple cluster with $K = 2$ different classes of applications. All PMs are homogeneous and can host up to $C = 4$ VMs. Each application class requests the allocation of single core VMs only. The interarrival time distribution of VM allocation requests (transitions α_k^+) is exponential with rate $\lambda_1 = 1$ for class 1 and $\lambda_2 = 3$ for class 2. Both rates are expressed as number of VM allocation requests per hour. While not particularly realistic, the values attributed to them allow to focus on the policies' contribution to system behavior. Deallocation (transitions α_k^-) rates per class are $\mu_1 = \lambda_1$ and $\mu_2 = \lambda_2$, respectively, expressed as number of VM deallocation requests per hour. For each policy, we consider up to $M = 5$ PMs. Since we want to test a fully utilized system, N_k is set according to Eq. 1 ($N_k = 4 \cdot M/2$).

Fig. 4 shows results for these parametrizations. The x-axis corresponds to the number of PMs composing the system, the y-axis to the percentage of idle hosts, in steady-state. According to Eq. 2, a higher percentage of unused PMs implies a higher energy efficiency. From Fig. 4 it is possible to see that the percentage of unused PMs is affected by different policies. The *ML-LL* policy clearly outperforms the others as the total number of PMs increases, reaching nearly 40% idle machines in steady state. Even with different values, all policies show a higher percentage of idle hosts as M increases with the exception of *LL-ML*. Here, the allocation on new VMs on the least loaded (possibly empty) PM, together with a deallocation strategy that hardly unloads a given PM completely, contributes to an increase of power consumption as M increases.

Using an approximate solution (#SolutionType APPROXIMATE_EVMDD) for the *Rnd-Rnd* policy, we raise the number of PMs up to $M = 9$ and use different *load factors*, that is, different values of $\rho_k = \lambda_k/\mu_k$. We only consider values of $\rho \leq 1$ to test a system whose load intensity is not requiring its full capacity. In such a situation, the probability of having idle servers is sufficient to obtain benefits from the application of energy saving strategies. The service rates are fixed to $\mu_1 = 1$ and $\mu_2 = 3$, while λ_1 and λ_2 are scaled to obtain different ρ values (for simplicity, we assume $\rho_1 = \rho_2$).

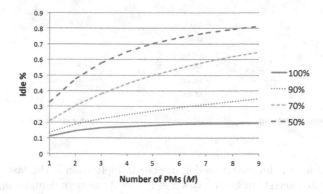

Fig. 5. Approximate results for the *Rnd-Rnd* policy under different load intensities

Fig. 5 shows the results. As expected, when ρ decreases, the probability to have idle machines increases. Unless noted otherwise, the rest of the paper assumes $\lambda = \mu$ to highlight the policy effects on the power preservation.

6.2 Modelling Different Types of VMs

For the second set of experiments we improve the characterization of different application classes. In particular, we allow different resource demands for each application, in terms of the number C of resources required by their VMs, thus providing a more accurate representation of the investigated system. The models in Fig. 2 and 3 are modified by adding a weight w_k to the arcs connecting places PM_m to transitions $g_{k,m}$, representing this higher resources demand by class. The same weight w_k is applied to the arcs from transitions $h_{k,m}$ to places PM_m, representing the deallocation of the same resource amount when a VM of that class is released. To satisfy the assumption of Eq. 1, the value of N for each class is now divided by the resources demands w_k, i.e., $N_k = M \cdot C / (K \cdot w_k)$.

Fig. 6 shows the results for the improved models. For these tests, we use $w_1 = 1$ and $w_2 = 2$ and investigate a different load balance: in Fig. 6a, $\lambda_1 = \mu_1 = 1$ while $\lambda_2 = \mu_2 = 3$ (class 2 is the fastest but also the most demanding in terms of resources); in Fig. 6b, $\lambda_1 = \mu_1 = 3$ while $\lambda_2 = \mu_2 = 1$ (class 2 is the slowest but still the most demanding one). The comparison of these two figures reveals an interesting behavior. For some but not all the cases, the resulting percentage of idle machines is the same, suggesting that policies are not influenced in the same way by different load balancing, revealing some that are more resilient to workload characteristics. In particular, *ML-ML* results in slightly more idle servers when class 2 is the slowest but most resource demanding, while *RR-Rnd* shows the opposite behavior. Another interesting fact can be observed when the system has 3 PMs: on the *ML-ML* and *ML-LL* curves there is a sort of "step". A possible reason is an uneven allocation of VMs that, for some states, prevents the allocation of a new 2-core VM on the most loaded server, requiring to allocate

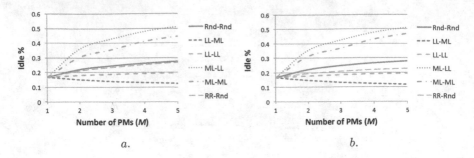

a.   *b.*

Fig. 6. Results for the improved model. *a*: class 2 application is the fastest; *b*: class 2 application is the slowest. In both, class 2 is the most resource demanding.

it on a (possibly) unused one, instead. This effect is presumably less evident as the number of PMs raises.

6.3   Performance Analysis and Energy Consumption

This section increases accuracy by including performance degradation of a PM when the number of running VMs on it increases. To this end, the firing of transition α_k^- is slowed by a factor 10 for each resource used. For the tests in this section, we focus on a specific parametrization of the model: we fix the number of PMs to $M = 4$, all providing the same number $C = 4$ of resources. $K = 2$ applications run in the system, each with different allocation and deallocation rates ($\lambda_1 = \mu_1 = 3$, $\lambda_2 = \mu_2 = 1$) and different resource demands ($w_1 = 1$, $w_2 = 2$). The maximum number of instances per application is $N_1 = 8$ and $N_2 = 4$, respectively. Fig. 7 shows, for each policy, the steady-state probability of having a given number of idle PMs. While the mean number of idle machines used in the previous sections is a useful estimate, we present here the distribution of this number per machine, to show more clearly how each policy affects energy consumption. For all cases in Fig. 7, the probability to have 3 or 4 unutilised PMs is negligible (less than 1%). Policies like *ML-LL* and *ML-ML* have the highest probability to leave idle 1 or 2 machines. Interestingly, *ML-LL* shows a lower probability to have no PMs idle than having only 1 in use.

The interference caused by multiple VMs active on the same PM can change the execution time of any batch process executed by that machine, especially when their combined workload pushes the utilization close to the maximum. Different allocation and deallocation policies can minimize or magnify this interference. We therefore introduce another peformance measure for the model: the *system running time* (R) representing the mean execution time of each allocated VM, considering their mutual interactions. The introduction of R helps relate energy consumption with system performance. This measure is equivalent to system response time for time-sharing systems and can be computed using the well known Little's Law ($N = X \cdot R$).

Fig. 8 shows results for different policies, where R is expressed in hours. *ML-ML* shows a behavior opposite to *LL-LL*: *ML-ML* allows a system to have

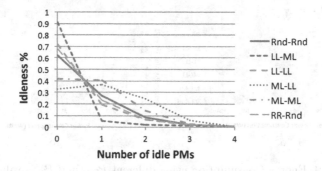

Fig. 7. Steady state probability to have a given number of idle PMs, per policy

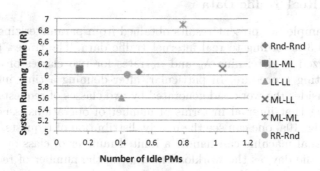

Fig. 8. Running time in hours computed using Little's Formula

more idle PMs, but at the cost of the worst system running time; *LL-LL*, while providing the best running time, requires more running PMs. Other policies provide similar R, but with very different energy consumption values.

In previous sections, we used Eq. 2 as a measure of energy consumption. To give a more accurate account of energy consumption per policy, we now apply Eq. 4. Fig. 9 shows results obtained by its use, for different values of P_{idle} and P_{busy}, as in [3]. The values used in Fig. 9a, $P_{idle} = 18$ Watts and $P_{busy} = 67$ Watts were measured on a laptop machine and may represent, in a virtualised datacenter, the newest servers on the market, with low consumption rates. In Fig. 9b, $P_{idle} = 70$ Watts and $P_{busy} = 160$ Watts values approximate instead measurements taken on a desktop i7 Asus machine. They may well represent the typical machine used as a server in a private cloud environment.

Different P_{idle} and P_{busy} values result in different power consumptions in the two figures, but they behave proportionally (the *LL-ML* always has the highest consumption). The only noticeable exceptions are *ML-LL* and *ML-ML*, as they tend to consolidate VMs over the fewest possible PMs. This strategy works best when P_{idle} is low, while it is less effective when it is high, as there is little gain in turning off a machine that does not consume much power when idle.

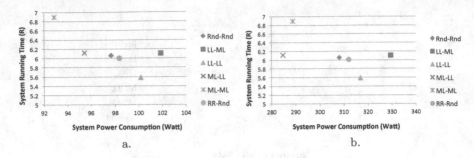

Fig. 9. Energy Consumption using different P_{idle} and P_{busy} values

6.4   Using Real Traffic Data

As a final example, we present results obtained from parameterizing the model in Fig. 2 and 3 according to real internet traffic data. The traces, available at [6], are analyzed to determine λ_k and μ_k rates for two classes of applications through a fitting procedure. In particular, after defining a time interval of 24 hours, we consider the workload generated by two classes of requests, GET and $POST$, during that interval in terms of number of operations issued every 15 minutes. In Fig. 10a, one can see the hourly distribution of requests. We assume that each virtual machine can handle a limited number of class operations per hour. During the day, as the workload fluctuates, the number of required VMs will vary. We set the maximum number of requests per hour handled by class 1 VMs at 1.25 million, while class 2 VMs can handle up to 100,000 requests per hour. In Fig. 10a, the main y-axis shows the thresholds for the first class of requests (GET) while the secondary y-axis shows the same for the second class (PUT). As it can also be seen from the figure's grid, we need a total of 18 VMs to handle the daily load, 9 for each class. From the traces it is possible to evaluate for how many 15 minute intervals a given number of VMs of a class is required to support the workload. We then set λ_k and μ_k dependent to the marking of p_k, to reproduce, on average, the traces' throughput. Unlike for other tests, we assume that there is no degradation in performance when a PM utilization increases and that, in this case, $\lambda_k(p_k) \neq \mu_k(p_k)$. This requires to measure energy efficiency using Eq. 2, as the mean number of idle machines in steady state. Moreover, all the VMs requested by classes are of the same type ($w_1 = w_2 = 1$). The maximum number of virtual machines required by one of the $K = 2$ classes considered is fixed at $N = 8$. This is because we assume that at least 1 VM per class is always on to handle the minimum load expected and is thus excluded from the model. Fig. 10b presents results for three different combinations of M and C able to allocate the total of 16 VMs required. By Eq. 1, $M \cdot C = 16$; thus, if the system has $M = 2$, then each PM machine has $C = 8$ cores; if $M = 4$, then $C = 4$; and, if $M = 8$, then $C = 2$.

The figure shows increasing probabilities to have idle PMs as M grows, as observed in Section 6.1. In this case, it is possible to compare how different

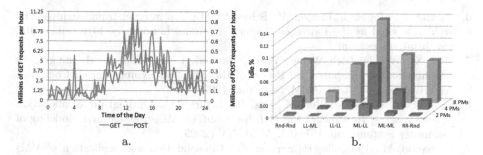

Fig. 10. a) Real traffic traces for two classes of requests. b) Comparing energy efficiency of different policies using real traffic data to set the model.

policies behave as PMs increase. Even if *ML-LL* provides the most idle PMs, policies like *LL-LL* and *Rnd-RR* seem to benefit more from a higher number of PMs, in particular showing, from $M = 4$ to $M = 8$, an 380% idleness increase.

7 Conclusion

We used Petri net models to investigate how different VM allocation and deallocation strategies impact energy consumption in a virtualised datacenter. We showed that different performance objectives, in terms of response time versus power consumption can be achieved by choosing the appropriate placement policy. Future implementation will include features like times and power costs for server startup and shutdown. To extend the realism of the model, more complex policies as well as different classes of resources, like storage and network, will also be included in addition to the CPU. Due to state-space explosion, such model extensions will likely require discrete-event simulation for their study.

References

1. Beloglazov, A., Buyya, R., Lee, Y.C., Zomaya, A., et al.: A taxonomy and survey of energy-efficient data centers and cloud computing systems. Advances in Computers **82**(2), 47–111 (2011)
2. Bennani, M.N., Menascé, D.A.: Resource allocation for autonomic data centers using analytic performance models. In: Autonomic Computing, ICAC 2005, pp. 229–240, June 2005
3. Cerotti, D., Gribaudo, M., Piazzolla, P., Pinciroli, R., Serazzi, G.: Multi-class queuing networks models for energy optimization. In: Proc. of 8th Int. Conf. Performance Evaluation Methodologies and Tools (2014)
4. Cerotti, D., Gribaudo, M., Piazzolla, P., Serazzi, G.: Matching performance objectives for open and closed workloads by consolidation and replication. Annals of Operations Research, pp. 1–24 (2014)
5. Ciardo, G., Jones III, R.L., Miner, A.S., Siminiceanu, R.I.: Logic and stochastic modeling with SMART. Performance Evaluation **63**(6), 578–608 (2006). Modelling Techniques and Tools for Computer Performance Evaluation

6. Danzig, P., Mogul, J., Paxson, V., Schwartz, M.: The internet traffic archive. The archive is sited at the Lawrence Berkeley National Laboratory. http://ita.ee.lbl. gov/html/traces.html
7. Fan, X., Wolf-Dietrich Weber, W.-D., Barroso, L.A.: Power provisioning for a warehouse-sized computer. In: Proce. of the 34th Int. Symposium on Computer Architecture, pp. 13–23. ACM, New York (2007)
8. Gandhi, A., Harchol-Balter, M., Das, R., Lefurgy, C.: Optimal power allocation in server farms. In: Proc. of the 11th Int. Conf. on Measurement and Modeling of Computer Systems, pp. 157–168. ACM, NY (2009)
9. Gribaudo, M., Piazzolla, P., Serazzi, G.: Consolidation and replication of VMs matching performance objectives. In: Al-Begain, K., Fiems, D., Vincent, J.-M. (eds.) ASMTA 2012. LNCS, vol. 7314, pp. 106–120. Springer, Heidelberg (2012)
10. Nathuji, R., Schwan, K.: Virtualpower: Coordinated power management in virtualized enterprise systems. SIGOPS Oper. Syst. Rev. **41**(6), 265–278 (2007)
11. Pinheiro, E., Bianchini, R., Carrera, E.V., Heath, T.: Load balancing and unbalancing for power and performance in cluster-based systems (2001). http://www2. ic.uff.br/julius/stre/pinheiro01load.pdf
12. Raghavendra, R., Ranganathan, P., Talwar, V., et al.: No "power" struggles: Coordinated multi-level power management for the data center. SIGARCH Comput. Archit. News **36**(1), 48–59 (2008)
13. Song, Y., Wang, H., Li, Y., et al.: Multi-tiered on-demand resource scheduling for VM-based data center. In: Proc. of the 9th Symposium on Cluster Computing and the Grid, pp. 148–155. IEEE (2009)
14. Watson, B.J., Marwah, M., Gmach, D., et al.: Probabilistic performance modeling of virtualized resource allocation. In: Proc. of the 7th int. Conference on Autonomic Computing, pp. 99–108. ACM, NY (2010)
15. Xu, X., Hu, H., Hu, N., Ying, W.: Cloud task and virtual machine allocation strategy in cloud computing environment. In: Lei, J., Wang, F.L., Li, M., Luo, Y. (eds.) NCIS 2012. CCIS, vol. 345, pp. 113–120. Springer, Heidelberg (2012)

On the Influence of High Priority Customers on a Generalized Processor Sharing Queue

Jasper Vanlerberghe$^{(\boxtimes)}$, Joris Walraevens, Tom Maertens, and Herwig Bruneel

Stochastic Modelling and Analysis of Communication Systems Research Group
(SMACS), Department of Telecommunications and Information Processing (TELIN),
Ghent University (UGent), Sint-Pietersnieuwstraat 41, B-9000 Gent, Belgium
{jpvlerbe,jw,tmaerten,hb}@telin.UGent.be

Abstract. In this paper, we study a hybrid scheduling mechanism in discrete-time. This mechanism combines the well-known Generalized Processor Sharing (GPS) scheduling with strict priority. We assume three customer classes with one class having strict priority over the other classes, whereby each customer requires a single slot of service. The latter share the remaining bandwith according to GPS. This kind of scheduling is used in practice for the scheduling of jobs on a processor and in Quality of Service modules of telecommunication network devices. First, we derive a functional equation of the joint probability generating function of the queue contents. To explicitly solve the functional equation, we introduce a power series in the weight parameter of GPS. Subsequently, an iterative procedure is presented to calculate consecutive coefficients of the power series. Lastly, the approximation resulting from a truncation of the power series is verified with simulation results. We also propose rational approximations. We argue that the approximation performs well and is extremely suited to study these systems and their sensitivity in their parameters (scheduling weights, arrival rates, loads ...). This method provides a fast way to observe the behaviour of such type of systems avoiding time-consuming simulations.

Keywords: Generalized Processor Sharing (GPS) · Priority · Queueing · Scheduling · Power series

1 Introduction

Numerous queueing systems in practice, have a high-priority bypass possibility. In this paper we study the influence of these high priority customers on a generalized processor sharing (GPS) queue. For instance, the processor of a computer system is shared by several jobs, whereby each class of jobs gets a time-share according to the weight of its class. However, the processor can also be interrupted, for hardware I/O for instance (i.e., the user pushes a key, requested data from the harddisk becomes available ...), these are in fact short high-priority jobs, bypassing the normal scheduling mechanism.

© Springer International Publishing Switzerland 2015
M. Gribaudo et al. (Eds.): ASMTA 2015, LNCS 9081, pp. 203–216, 2015.
DOI: 10.1007/978-3-319-18579-8_15

An example from telecommunications is DiffServ [9]. DiffServ is short for Differentiated Services and is an architecture designed to deliver a different Quality of Service (QoS) grade to various services in telecommunication networks. It defines an Expedited Forwarding (EF) class of packets next to the Assured Forwarding (AF) class. EF packets have essentially high priority and are thus given strict priority over all other packets. The AF class of packets is divided into subclasses, and the scheduling amongst the subclasses is a GPS-based scheduling.

Cisco implemented this kind of scheduling mechanism in some of its gigabit switch routers. The brand names used are IP Realtime Transport Protocol (RTP) Priority and Low Latency Queueing (LLQ); both are based on a mixture of GPS-like scheduling with priority bypassing. They differ in the type of traffic they support, i.e., UDP vs TCP.

As a result of its practical application, this model also attracted attention from the research community, where it is frequently referred to as PQ-GPS. Jin et al. [4,5] studied PQ-GPS under long-range dependent traffic by using a flow decomposition approach dividing the system into single-server single-queue (SSSQ) systems. They obtain analytical upper and lower bounds. Parveen [12] used the same SSSQ approach to study a system containing both long-range and short-range dependent traffic. After the single queue decomposition he however uses another technique resulting in a single approximation, as opposed to an upper and lower bound. Lastly, we mention Wang et al. [20] who studied a finite hybrid queueing model using PQ and Weighted Fair Queueing (WFQ). As WFQ is known to be a good approximation for GPS, it is also of interest here. Drawing up a Markov chain for the system and solving it for the steady-state probability, they conclude with a sensitivity analysis for the parameters of the system.

Next to studying hybrid scheduling models, most of the attention has gone to both individual models, i.e., either priority queueing or generalized processor sharing models. Priority queueing was, for instance, studied in [3,6,13,15,18,19]. Whereas, GPS was analyzed in [7,8,10,11,17,21].

In this paper, we analyze a hybrid priority-GPS scheduling algorithm. We construct a functional equation for the probability generating function (pgf) of the queue contents in steady state. Subsequently, we develop an iterative procedure to calculate the coefficients of the power series of this pgf, whereby the power series is constructed in the GPS-weight. Due to practical restrictions, we use the truncated power series to construct approximations. Lastly, we evaluate the approximations using simulation results.

2   Mathematical Model

We consider a discrete-time (i.e., time is assumed to be slotted) queueing system with three queues of infinite capacity and one transmission channel. Three classes of customers, named 1, 2 and 3, arrive to the system. Customers of class 1 have strict priority over the other customers. Consequently, the server always serves class 1 as long as this class is backlogged. If class 1 is not backlogged, class 2 and 3 customers are served according to a discrete-time implementation

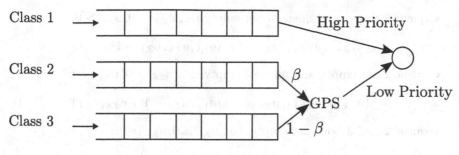

Fig. 1. Model

of GPS. As such, the server serves a class 2 customer with probability β and a class 3 customer with probability (w.p.) $1-\beta$, if both classes are backlogged. The weight parameter of the GPS scheduling is thus β and can be used to divide the bandwith among customers of class 2 and 3. Within each queue, the customers are served in FIFO order. This model is depicted in Fig. 1.

The number of arrivals of class j ($j = 1, 2, 3$) in slot k is denoted by $a_{j,k}$, where we assume $\{a_{j,k}, k > 0\}$ forms a sequence of independent and identically distributed random variables. The joint pgf of the arrivals of all classes is denoted as $A(z_1, z_2, z_3) \triangleq E[z_1^{a_{1,k}} z_2^{a_{2,k}} z_3^{a_{3,k}}]$. Furthermore, we define λ_j as the mean number of arrivals in queue j and λ_T as the mean total number of arrivals to the queueing system per time slot. Every customer requires a single slot of service. This means that the load ρ (i.e., the mean number of slots of work arriving to the system per slot) equals λ_T; subsequently, the stability condition for this queueing system is $\lambda_T < 1$.

In the next sections, we study the stationary distribution of the queue content in each of the queues. Therefore, we define $u_{j,k}$ as the queue content in queue j at the beginning of slot k and $U_k(z_1, z_2, z_3) \triangleq E[z_1^{u_{1,k}} z_2^{u_{2,k}} z_3^{u_{3,k}}]$ as the joint pgf of the queue content at the beginning of slot k. The stationary distribution is then $U(z_1, z_2, z_3) = \lim_{k\to\infty} U_k(z_1, z_2, z_3)$.

3   The Functional Equation

Let us first establish the system equations, relating $(u_{1,k}, u_{2,k}, u_{3,k})$ and $(u_{1,k+1}, u_{2,k+1}, u_{3,k+1})$, i.e., the state of the system at the beginning of slot k and the state of the system at slot $k + 1$. We split the equations into several (sub)cases:

- **All queues empty,** i.e., $u_{j,k} = 0, j = 1, 2, 3$:

$$(u_{1,k+1}, u_{2,k+1}, u_{3,k+1}) = (a_{1,k}, a_{2,k}, a_{3,k}) \tag{1}$$

- **Queue 1 not empty,** i.e., $u_{1,k} > 0$:

$$(u_{1,k+1}, u_{2,k+1}, u_{3,k+1}) = (u_{1,k} - 1 + a_{1,k}, u_{2,k} + a_{2,k}, u_{3,k} + a_{3,k}) \tag{2}$$

- **Queue 1 empty,** i.e., $u_{1,k} = 0$:

- queue 2 empty and queue 3 not empty i.e., $u_{2,k} = 0, u_{3,k} > 0$:

$$(u_{1,k+1}, u_{2,k+1}, u_{3,k+1}) = (a_{1,k}, a_{2,k}, u_{3,k} - 1 + a_{3,k}) \tag{3}$$

- queue 2 not empty and queue 3 empty, i.e., $u_{2,k} > 0, u_{3,k} = 0$:

$$(u_{1,k+1}, u_{2,k+1}, u_{3,k+1}) = (a_{1,k}, u_{2,k} - 1 + a_{2,k}, a_{3,k}) \tag{4}$$

- queue 2 and 3 both not empty, i.e., $u_{2,k} > 0, u_{3,k} > 0$:

$$(u_{1,k+1}, u_{2,k+1}, u_{3,k+1}) = \begin{cases} (a_{1,k}, u_{2,k} - 1 + a_{2,k}, u_{3,k} + a_{3,k}) \text{ w.p. } \beta \\ (a_{1,k}, u_{2,k} + a_{2,k}, u_{3,k} - 1 + a_{3,k}) \text{ w.p. } 1 - \beta \end{cases} \tag{5}$$

From these systems equations, we construct a relation between the pgfs $U_k(z_1, z_2, z_3)$ and $U_{k+1}(z_1, z_2, z_3)$:

$$\begin{aligned} U_{k+1}(z_1, z_2, z_3) = A(z_1, z_2, z_3)\bigg(& U_k(0,0,0) \\ & + \frac{1}{z_1}\big(U_k(z_1, z_2, z_3) - U_k(0, z_2, z_3)\big) \\ & + \frac{1}{z_3}\big(U_k(0, 0, z_3) - U_k(0,0,0)\big) \\ & + \frac{1}{z_2}\big(U_k(0, z_2, 0) - U_k(0,0,0)\big) \\ & + \left(\frac{\beta}{z_2} + \frac{1-\beta}{z_3}\right)\big(U_k(0, z_2, z_3) \\ & - U_k(0,0,z_3) - U_k(0, z_2, 0) + U_k(0,0,0)\big)\bigg). \end{aligned} \tag{6}$$

In steady state, both U_k and U_{k+1} are equal. We denote $U(z_1, z_2, z_3) \triangleq \lim_{k\to\infty} U_k(z_1, z_2, z_3) = \lim_{k\to\infty} U_{k+1}(z_1, z_2, z_3)$ as the pgf of the queue content in steady state. By letting $k \to \infty$ in Equation (6) and solving the result for $U(z_1, z_2, z_3)$, we retrieve the following functional equation for $U(z_1, z_2, z_3)$:

$$U(z_1, z_2, z_3) = \frac{A(z_1, z_2, z_3)\begin{cases} \big(z_2(z_1 - z_3) + \beta z_1(z_3 - z_2)\big)U(0, z_2, z_3) \\ +(1 - \beta)z_1(z_3 - z_2)U(0, z_2, 0) \\ -\beta z_1(z_3 - z_2)\big)U(0, 0, z_3) \\ +\big(z_1 z_3(z_2 - 1) + \beta z_1(z_3 - z_2)U(0,0,0) \end{cases}}{z_2 z_3(z_1 - A(z_1, z_2, z_3))} \tag{7}$$

This functional equation still contains some unknowns that need to be determined to obtain full knowledge of the statistical distribution of the queue length.

Therefore, we need to calculate the unknown boundary functions $U(0, z_2, z_3)$, $U(0, z_2, 0)$, $U(0, 0, z_3)$ and $U(0, 0, 0)$. This last unknown is easily found as $U(0, 0, 0)$ and equals the probability that the system is empty in steady state ($u_1 = u_2 = u_3 = 0$). In queueing theory, this is a well-known result and is equal to $1 - \lambda_T$. For ease of notation, however, we will only do this substitution after eliminating the other boundary functions.

4   The Power Series Approximation

To eliminate the boundary functions, we write $U(z_1, z_2, z_3)$ as a power series in β, where we assume $U(z_1, z_2, z_3)$ is analytic in a neighborhood of $\beta = 0$. This approach was also used in [17] to analyze a two-queue GPS system. We write:

$$U(z_1, z_2, z_3) = \sum_{m=0}^{\infty} V_m(z_1, z_2, z_3) \beta^m. \tag{8}$$

In the remainder of this section, we use this power series and the functional equation from the previous section to derive an iterative procedure to calculate V_m from V_{m-1}.

4.1   Eliminating $V_m(0, 0, z_3)$

The first step is to replace $U(z_1, z_2, z_3)$ by its power series in (7). Subsequently, we can equate the coefficients of β^m on the right and left hand side of Equation (7). For the coefficient of $\beta^m, m \geq 0$, this yields

$$(z_2 z_3 (z_1 - A(z_1, z_2, z_3))) V_m(z_1, z_2, z_3) \tag{9}$$
$$= A(z_1, z_2, z_3) \Big[z_1 (z_3 - z_2) \Big(P_{m-1}(z_2, z_3) + V_m(0, z_2, 0) + V_{m-1}(0, 0, 0) \Big)$$
$$+ z_2 (z_1 - z_3) V_m(0, z_2, z_3) + z_1 z_3 (z_2 - 1) V_m(0, 0, 0) \Big],$$

where we defined $V_{-1}(z_1, z_2, z_3) \triangleq 0$ and $P_m(z_2, z_3) = V_m(0, z_2, z_3) - V_m(0, z_2, 0)$ $- V_m(0, 0, z_3)$ and thus $P_{-1}(z_2, z_3) = 0$. Looking closely at Equation (9), we can see that only two of the remaining unknown boundary functions $V_m(0, z_2, 0)$ and $V_m(0, z_2, z_3)$ are needed to calculate $V_m(z_1, z_2, z_3)$, assuming $V_{m-1}(z_1, z_2, z_3)$ is known. By introducing the power series we effectively eliminated one of the unknown boundary functions.

4.2   Eliminating $V_m(0, z_2, z_3)$

By using a generalization of Rouchè's theorem [1], we can prove that $z_1 -$ $A(z_1, z_2, z_3)$ has one zero in the unit disk of z_1 for an arbitrary z_2 and z_3 in the unit disk. We denote this zero by $Y_{2,3}(z_2, z_3)$ and it is thus implicitly defined as $Y_{2,3}(z_2, z_3) = A(Y_{2,3}(z_2, z_3), z_2, z_3)$, with $|Y_{2,3}(z_2, z_3)| < 1$. As the left hand side of Equation (9) is zero for $z_1 = Y_{2,3}(z_2, z_3)$ and $V_m(z_1, z_2, z_3)$ remains finite in the unit circle, the right hand side should also equal zero. This leads to

$$z_2(z_3 - Y_{2,3}(z_2, z_3))V_m(0, z_2, z_3)$$
$$= Y_{2,3}(z_2, z_3)(z_3 - z_2)\Big(P_{m-1}(z_2, z_3) + V_m(0, z_2, 0) + V_{m-1}(0, 0, 0)\Big)$$
$$+ Y_{2,3}(z_2, z_3)z_3(z_2 - 1)V_m(0, 0, 0). \tag{10}$$

4.3 Eliminating $V_m(0, z_2, 0)$

We can prove that $Y_{2,3}(z_2, z_3)$ is the pgf of a random variable of this system, see [18] for a similar example. Then by again using Rouchè's theorem, we can prove that $z_3 - Y_{2,3}(z_2, z_3)$ has one zero in the unit disk of z_3 for an arbitrary z_2 in the unit disk. We denote this zero by $Y_2(z_2)$ and it is thus implicitly defined as $Y_2(z_2) = Y_{2,3}(z_2, Y_2(z_2)) = A(Y_2(z_2), z_2, Y_2(z_2))$, with $|Y_2(z_2)| < 1$. As the left hand side of Equation (10) is zero for $z_3 = Y_2(z_2)$ and $V_m(0, z_2, z_3)$ remains finite in the unit circle, the right hand side should also equal zero. This yields

$$V_m(0, z_2, 0) = -P_{m-1}(z_2, Y_2(z_2)) - V_{m-1}(0, 0, 0) + \frac{Y_2(z_2)(z_2 - 1)V_m(0, 0, 0)}{z_2 - Y_2(z_2)}. \tag{11}$$

Feeding this result back into Equation (10), we get that

$$z_2(z_3 - Y_{2,3}(z_2, z_3))V_m(0, z_2, z_3)$$
$$= Y_{2,3}(z_2, z_3)(z_3 - z_2)\Big(Q_{m-1}(z_2, z_3) + \frac{Y_2(z_2)(z_2 - 1)V_m(0, 0, 0)}{z_2 - Y_2(z_2)}\Big), \tag{12}$$

with

$$Q_m(z_2, z_3) = P_m(z_2, z_3) - P_m(z_2, Y_2(z_2)) \tag{13}$$
$$= V_m(0, z_2, z_3) - V_m(0, z_2, Y_2(z_2)) - V_m(0, 0, z_3) + V_m(0, 0, Y_2(z_2)).$$

Lastly, as $U(0, 0, 0) = 1 - \lambda_T$ (shown before), we know that $V_0(0, 0, 0) = 1 - \lambda_T$ and $V_m(0, 0, 0) = 0$ for $m > 0$.

So by introducing the power series notation and the two implicitly defined functions $Y_{2,3}$ and Y_2, we found a solution for the boundary functions. Substituting, these solutions in Equation (9), we get (with $m > 0$) that

$$V_0(z_1, z_2, z_3) = \frac{(1 - \lambda_T)A(z_1, z_2, z_3)(z_2 - 1)(z_3 - Y_2(z_2))(z_1 - Y_{2,3}(z_2, z_3))}{(z_2 - Y_2(z_2))(z_3 - Y_{2,3}(z_2, z_3))(z_1 - A(z_1, z_2, z_3))}, \tag{14}$$

$$V_m(z_1, z_2, z_3) = \frac{A(z_1, z_2, z_3)(z_3 - z_2)Q_{m-1}(z_2, z_3)(z_1 - Y_{2,3}(z_2, z_3))}{z_2(z_3 - Y_{2,3}(z_2, z_3))(z_1 - A(z_1, z_2, z_3))}. \tag{15}$$

As a result, starting from V_0, V_m can be calculated from V_{m-1}. This concludes the iterative calculation procedure of $U(z_1, z_2, z_3)$.

As a test of our analysis, suppose we would want to study the joint probability generating function of u_1 and $u_2 + u_3$. We can do this by replacing both z_2 and z_3 by z, as $E[z_1^{u_1} z^{u_2+u_3}] = U(z_1, z, z)$. We subsequently get:

$$V_0(z_1, z, z) = \frac{(1 - \lambda_T) A(z_1, z, z)(z - 1)(z_1 - Y_{2,3}(z, z))}{(z - Y_{2,3}(z, z))(z_1 - A(z_1, z, z))}, \tag{16}$$

$$V_m(z_1, z, z) = 0. \tag{17}$$

As V_m equals zero for $m > 0$, $U(z_1, z, z) = V_0(z_1, z, z)$ and the pgf is independent of β, as expected. The result we get, is the pgf for a priority queueing system with 2 queues as can be found in [19]. This confirms our result.

5   Approximations of Performance Measures

In the previous section, we derived an iterative algorithm to calculate the joint pgf $U(z_1, z_2, z_3)$ of the queue content. More practical performance measures of the system, however, would for instance be the mean length of each of the three queues. These can be calculated from the power-series form of the pgf

$$E[u_j] = \left. \frac{\partial U(z_1, z_2, z_3)}{\partial z_j} \right|_{z_1 = z_2 = z_3 = 1}$$

$$= \sum_{m=0}^{\infty} \beta^m \left. \frac{\partial V_m(z_1, z_2, z_3)}{\partial z_j} \right|_{z_1 = z_2 = z_3 = 1}. \tag{18}$$

We showed earlier that $V_m(z_1, 1, 1) = 0$ for $m > 0$, so $E[u_1] = V_0(1, 1, 1)$ is independent of β. This is of course expected, as the length of the high-priority queue should not depend on the scheduling of the packets of the lower priority queues.

A second conclusion follows from the fact that in the work conserving system presented here, the total backlog is a constant. This constant $E[u_T]$ is independent of β. As a result, we get:

$$E[u_T] = E[u_1] + E[u_2] + E[u_3], \tag{19}$$

$$E[u_T] - E[u_1] = E[u_2] + E[u_3] \tag{20}$$

$$= \sum_{m=0}^{\infty} \beta^m \left. \frac{\partial V_m(1, z_2, 1)}{\partial z_2} \right|_{z_2=1} + \sum_{m=0}^{\infty} \beta^m \left. \frac{\partial V_m(1, 1, z_3)}{\partial z_3} \right|_{z_3=1}. \tag{21}$$

The terms in the left hand side are constants, while $E[u_2]$ and $E[u_3]$ on the right hand side of the equation are a function of β, as can be seen from Equation (18). Subsequently, this means that for $m > 0$:

$$\left. \frac{\partial V_m(1, z_2, 1)}{\partial z_2} \right|_{z_2=1} = - \left. \frac{\partial V_m(1, 1, z_3)}{\partial z_3} \right|_{z_3=1}. \tag{22}$$

This result can significantly help speed up calculations, as we only need to calculate one of those derivatives.

With these results, we are able to calculate the exact mean queue lengths,
or at least to an arbitrary precision. This is however only theoretically possible.
In practice, the calculation of V_m is far from straightforward. The calculation
of Q_{m-1} in (15) involves $V_m(0, z_2, Y_2(z_2))$ and $V_m(0, 0, Y_2(z_2))$, for which several applications of l'Hopital's rule are needed. The differentiation in l'Hopital's
rule leads to very large expressions, quickly becoming infeasable for current
computers. Calculating the mean queue length involves another differentiation
and evaluation in 1 for all $z_j, j = 1..3$, leading to several more applications of
l'Hopital's rule.

We, however, have another trick up our sleeve. We can also calculate the
power series in $\beta = 1$ leading to:

$$U(z_1, z_2, z_3) = \sum_{m=0}^{\infty} (1 - \beta)^m \tilde{V}_m(z_1, z_2, z_3) \tag{23}$$

So, because of the symmetry in the system, $\tilde{V}_m$ can be calculated from V_m,
whereby class 3 customers are sent to queue 2 and class 2 customers to queue 3.
In particular, $\tilde{V}_m$ can be calculated from Equation (15) with $A(z_1, z_2, z_3)$ replaced
by $A(z_1, z_3, z_2)$. Subsequently, the mean lengths of queues 2 and 3 can be calculated as

$$E[u_j] = \sum_{m=0}^{\infty} \frac{\partial \tilde{V}_m(z_1, z_2, z_3)}{\partial z_{5-j}} \bigg|_{z_1 = z_2 = z_3 = 1}, \qquad j = 2, 3. \tag{24}$$

Basically, in practice we can calculate the first M terms of the power series
of $E[u_2]$ and $E[u_3]$, either in $\beta = 0$ or in $\beta = 1$, from the functions V_0 up to
V_M. With these values we can construct approximations. We opt to approximate
$E[u_2]$ and $E[u_3]$ by rational functions (Padè approximants) of the form

$$[L/N]_{E[u_j]}(\beta) = \frac{\sum_{l=0}^{L} v_{j,l} \beta^l}{\sum_{n=0}^{N} w_{j,n} \beta^n}, \tag{25}$$

whereby the coefficients $v_{j,l}$ and $w_{j,n}$ should be chosen such that the derivatives of $[L/N]_{E[u_j]}(\beta)$ in either 0 or 1 match the values obtained before. For
$[L/N]_{E[u_j]}(\beta)$ to be unique, we need a normalization. Therefore, we choose
$w_{j,0} = 1$. As we have $2(M + 1)$ datapoints and $L + N + 1$ coefficients in
$[L/N]_{E[u_j]}(\beta)$, we need to choose L and N such that $L + N = 2M + 1$.

The Padè approximants can introduce difficulties as the denominator can
introduce poles for $\beta \in [0, 1]$. Furthermore, the result could be non-monotone;
however, the mean queue length of class 2 (class 3) should decrease (increase)
in β. Lastly, the performance of each approximant is different and varies with
the parameters of the arrival process, so it is unclear which one performs best
beforehand (see also the numerical examples in the next section). These problems
are identical to the ones in [16], the solution presented therein can also be used

here to overcome these problems. This solution (in short) consists of disregarding the unfeasible approximants and averaging the remaining ones. As to keep this text self-contained and simple, we will restrict the discussion here to the Padè approximants (and in the remainder do not use the solution from [16].

6   Numerical Examples

In this section, we will compare our power series approximation for the mean queue length with simulation results. As the mean queue length for class 1 is not influenced by the other queues and could also easily be calculated from results for single-class FCFS queueing, we will not discuss it here. Furthermore, we only analyze queue 2, as the system is work conserving, results for queue 3 follow easily from (19).

We will use an arrival process with a joint pgf of the number of arrivals of the three classes of the form

$$A(z_1, z_2, z_3) = \left(1 + \frac{\lambda_1}{16}(z_1 - 1) + \frac{\lambda_2}{16}(z_2 - 1) + \frac{\lambda_3}{16}(z_3 - 1)\right)^{16}, \quad (26)$$

where λ_j is the arrival rate of class j customers (as defined earlier). Furthermore, we define $\alpha_1 = \frac{\lambda_1}{\lambda_T}$ and $\alpha_2 = \frac{\lambda_2}{\lambda_T}$ as the fraction of class 1 and class 2 customers, respectively.

For the simulation results in this section, we have used Monte-Carlo simulations over 10^7 slots. This high number of slots is enough to eliminate bias from the transient phase. Additionally, each simulation uses exactly the same sequence of arrivals and decision variables, to minimize the variance between simulations for different parameters of the system. This is the well-known technique of the common random numbers[2, 14].

In Fig. 2, we show the mean length of queue 2 as a function of the weight β, with $\lambda_T = 0.9, \alpha_1 = 0.1$, and $\alpha_2 = 0.1$. The figure shows curves of the simulation result and the Padè approximants without poles. We can see that for these parameters the [2/3] Padè approximant is very accurate.

Secondly, we observe that the approximations perform best close to $\beta = 0$ and $\beta = 1$. This is expected as the available information is exactly the value up to the M-th order derivative in these points (in this case $M = 2$). Subsequently, the approximants are constructed to match this information, thus performing well near $\beta = 0$ and $\beta = 1$.

In our second numerical example, we study the influence of the amount of high-priority (i.e., class 1) customers. We keep the total load $\lambda_T = 0.9$ fixed and $\lambda_2 = \lambda_3$, while increasing α_1 from 0.1 to 0.6. The mean queue-2 length is depicted in Fig. 3 on the left, showing both the simulation results and the best performing Padè approximant. We can see that the performance of the approximation is still accurate though slightly deteriorates as α_1 decreases, this results from the choice of the approximant. For this graph, we chose the [3/2] approximant, which on average performs best for these curves, but for smaller α_1 the [2/3] approximant is actually better. Furthermore for $\beta = 1$, i.e., when the

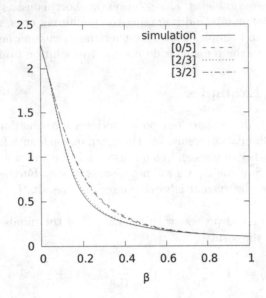

Fig. 2. Mean queue-2 length: comparison between simulation and Padè approximants

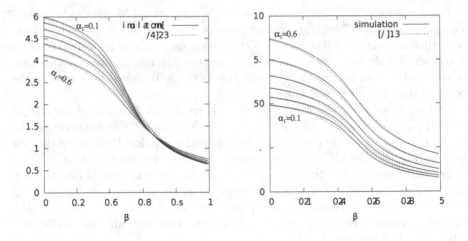

Fig. 3. Mean queue-2 length (left) and mean queue-2 delay (right): effect of increasing fraction of class 1 customers

queueing system is effectively a strict priority system with class 1 having highest priority, class 2 medium priority and class 3 low priority, higher α_1 barely makes a difference. This is mainly because there are few class 2 customers in the system as α_2 decreases from 0.1 to 0.056. On the other end for $\beta = 0$, we have a strict priority queueing system with class 1 high priority, class 3 medium priority and

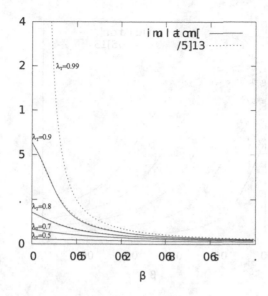

Fig. 4. Mean queue-2 length: effect of increasing total load

class 2 low priority. As class 2 is the lowest on the priority ladder, the influence of the bypassing (higher priority) class 3 and class 1 customers is greater. With α_2 small, however, queueing rarely happens and the influence is rather small.

Using Little's theorem, we also calculated the mean class-2 delay, it is depicted in Fig. 3 on the right. We saw before that as α_1 increases the mean queue-2 length decreases, mainly because α_2 decreases (we keep the total load and ratio between class-2 and 3 packets fixed). As we can see from the graph of the delay, for an increasing amount of high priority packets the class-2 packets have a larger delay. There are thus less class-2 packets in the system but they stay there longer.

In Fig. 4, we show $E[u_2]$ as a function of β for different values of the total load λ_T, with $\alpha_1 = \alpha_2 = 0.1$ fixed. As the load in the system increases, we observe the queue-2 length increases as well. This is a classical queueing result: a higher load always leads to higher congestion. As in the previous example (and for the same reason), we can see the effect at $\beta = 1$ is barely visible as opposed to at $\beta = 0$. Furthermore, we see that approximation is close to the simulated result. For $\lambda_T = 0.99$, we only depicted the approximation. Simulations over 10^7 slots do not converge for this high load, as the event of the system being empty becomes very rare.

Lastly, we look at the influence of the amount of class-2 customers while keeping the total load and the amount of high-priority packets constant. The results are depicted in Fig. 5 for $\lambda_T = 0.9, \alpha_1 = 0.1$ and α_2 ranging from 0.1 to 0.5. As the amount of class-2 packets increases the queue length increases, which was to be expected. Another observation is that the performance of the approximation deteriorates. In Fig. 5, we chose to show the [2/3] approximant. This is, however, not the best approximation for every parameter combination.

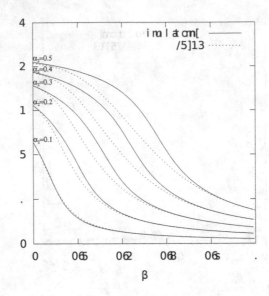

Fig. 5. Mean queue-2 length: effect of increasing fraction of class 2 customers

For instance, for $\alpha_2 = 0.5$ Padè approximant [3/2] is the best one. However, even if we compare every simulation with the best fitting approximant, the performance still deteriorates.

7 Conclusions

In this paper, we derived an analytical method to calculate the joint probability generating function of a three-class queueing system with a hybrid GPS-priority scheduling. The iterative algorithm leads to solutions with arbitrary precision in theory. Unfortunately, in practice, we are limited by the capabilities of current computers in the derivation of performance measures. Using Padè approximants, we have presented a method to use partial information to construct approximations. These approximations were compared with results from simulation and prove to work well. As a result, this power series approximation leads to a very efficient method to study these kind of systems for the whole parameter space, avoiding very time and resource consuming simulations.

Acknowledgments. This research has been co-funded by the Interuniversity Attraction Poles (IAP) Programme initiated by the Belgian Science Policy Office.

References

1. Adan, I.J., Van Leeuwaarden, J., Winands, E.M.: On the application of Rouché's theorem in queueing theory. Operations Research Letters **34**(3), 355–360 (2006)

2. Asmussen, S., Glynn, P.W.: Stochastic Simulation: Algorithms and Analysis: Algorithms and Analysis, vol. 57. Springer (2007)
3. Choi, B., Choi, D., Lee, Y., Sung, D.: Priority queueing system with fixed-length packet-train arrivals. IEE Proceedings-Communications 145(5), 331–336 (1998)
4. Jin, X., Min, G.: Analytical modelling of hybrid PQ-GPS scheduling systems under long-range dependent traffic. In: 21st International Conference on Advanced Information Networking and Applications, 2007, AINA 2007. pp. 1006–1013. IEEE (2007)
5. Jin, X., Min, G.: Performance modelling of hybrid PQ-GPS systems under long-range dependent network traffic. IEEE Communications Letters 11(5), 446–448 (2007)
6. Kim, K., Chae, K.C.: Discrete-time queues with discretionary priorities. European Journal of Operational Research 200(2), 473–485 (2010)
7. Lee, J.Y., Kim, S., Kim, D., Sung, D.K.: Bandwidth optimization for internet traffic in generalized processor sharing servers. IEEE Transactions on Parallel and Distributed Systems 16(4), 324–334 (2005)
8. Lieshout, P., Mandjes, M.: Generalized processor sharing: Characterization of the admissible region and selection of optimal weights. Computers & Operations Research 35(8), 2497–2519 (2008)
9. Nichols, K., Blake, S., Baker, F., Black, D.: Definition of the differentiated services field (DS field) in the IPv4 and IPv6 headers. RFC 2474 (Proposed Standard) (dec 1998). http://www.ietf.org/rfc/rfc2474.txt, updated by RFCs 3168, 3260
10. Parekh, A.K., Gallager, R.G.: A generalized processor sharing approach to flow control in integrated services networks: the single-node case. IEEE/ACM Transactions on Networking (TON) 1(3), 344–357 (1993)
11. Parekh, A.K., Gallagher, R.G.: A generalized processor sharing approach to flow control in integrated services networks: the multiple node case. IEEE/ACM Transactions on Networking (TON) 2(2), 137–150 (1994)
12. Parveen, A.S.: A survey of an integrated scheduling scheme with long-range and short-range dependent traffic. International Journal of Engineering Sciences & Research Technology 3(1), 430–439 (2014)
13. Smith, P.J., Firag, A., Dmochowski, P.A., Shafi, M.: Analysis of the M/M/N/N queue with two types of arrival process: Applications to future mobile radio systems. Journal of Applied Mathematics 2012 (2012)
14. Spall, J.C.: Introduction to stochastic search and optimization: estimation, simulation, and control, vol. 65. John Wiley & Sons (2005)
15. Takine, T., Sengupta, B., Hasegawa, T.: An analysis of a discrete-time queue for broadband ISDN with priorities among traffic classes. IEEE Transactions on Communications 42(234), 1837–1845 (1994)
16. Vanlerberghe, J., Walraevens, J., Maertens, T., Bruneel, H.: Approximating the optimal weights for discrete-time generalized processor sharing. In: Networking Conference, 2014 IFIP, pp. 1–9. IEEE (2014)
17. Walraevens, J., van Leeuwaarden, J., Boxma, O.: Power series approximations for two-class generalized processor sharing systems. Queueing systems 66(2), 107–130 (2010)
18. Walraevens, J., Steyaert, B., Bruneel, H.: Delay characteristics in discrete-time GI-G-1 queues with non-preemptive priority queueing discipline. Performance Evaluation 50(1), 53–75 (2002)

19. Walraevens, J., Steyaert, B., Bruneel, H.: Performance analysis of a single-server ATM queue with a priority scheduling. Computers & Operations Research **30**(12), 1807–1829 (2003)
20. Wang, L., Min, G., Kouvatsos, D.D., Jin, X.: Analytical modeling of an integrated priority and WFQ scheduling scheme in multi-service networks. Computer Communications **33**, S93–S101 (2010)
21. Zhang, Z.L., Towsley, D., Kurose, J.: Statistical analysis of the generalized processor sharing scheduling discipline. IEEE Journal on Selected Areas in Communications **13**(6), 1071–1080 (1995)

Author Index

Printed in the United States
By Bookmasters